Florida and Sea Level Rise

MODELS DESCRIBING THE SEA LEVEL RISE IN KEY WEST, FLORIDA

KARM-ERVIN JEAN

DEDICATION

I dedicate this manuscript to my God, my wife, and my kids. Its achievement would not have been possible without their love and great support.

Karm-Ervin Jean, MS Statistics

ACKNOWLEDGMENTS

I would like to thank Dr. B. M. Golam Kibria and Dr. Sneh Gulati for their indescribable guidance throughout my journey as both undergraduate and graduate student at Florida International University. Also, I would like to express my gratitude toward Dr. Harold R. Wanless from the Department of Geological Sciences at the University of Miami for providing me with the necessary resources for the accomplishment of this work.

Indisputably, none of the people mentioned above may be held responsible for any appearing errors in this book, independently made by my passion.

TABLE OF CONTENTS

CHAPTER	PAGE

LIST OF FIGURES

LIST OF TABLES

TABLE **PAGE**

Chapter 1

Introduction

There was a time when the mention of Florida Keys conjured up images of beautiful beaches, snorkeling, diving and just wonderful relaxing fun filled days. Climate change and sea level rise was not on anyone's radar screen. That now has changed. When we hear the name "Key West", we do not think "Margaritaville" but rather "climate change." Scientists believe that the sea-level in Key West will reach between 9 to 24 inches by 2060 (Strauss et al. 2012). The low lying coral island is at a risk of disappearing since the city's sea level rise has started to rapidly accelerate. Certainly this is not only true for Key West, but also most Floridian cities.

Intriguingly, Florida is rated number one among the states with the largest total population living on land less than 4 feet above high tide (Strauss et al. 2012). To make matters worse, our soil is porous making traditional methods of protecting against sea-level rise (sea-walls) useless since water seeps through the soil. Most of South Florida is defenseless against the devastating effects of sea level rise, and may be wiped out from the map.

According to the Florida Department of Community Affairs, Division of Community Planning, *"Communities that are subject to sea level rise may experience an increase in coastal vulnerability. Impacts to communities may include:*

- *Increased flooding and drainage problems,*
- *Destruction of natural resource habitats,*
- *Higher storms surge, increased evacuation areas and evacuation time frames,*
- *Increased shoreline erosion,*
- *Saltwater Intrusion, and*
- *Loss of infrastructure and existing development"*

What contribute to the sea level rise?

According to scientists, the volume of water in the ocean is primarily impacted by thermal expansion, melting ice from the poles, and rainwater. Thermal expansion due to only hot temperature contributes to at most 30% of the sea level rise (Cazenave and Llovel 2010). Melted ice from Greenland and Antarctic contributes to more than 55%. Melting from Ice sheets may dominate sea level rise in the 21st century (Rignot et al. 2011). Other influential factors for sea level rise of the ocean are salinity and atmospheric circulation.

Overall, the sea level rise is affected by "global warming" which does two things: heating up the atmosphere and causing the ocean to expand. It has direct impact on the polar ice sheets causing considerable amount of ice to melt into the sea. For the past few years, Key West was the center stage of many tropical storms and heavy floods, as reported in the news. Sadly, global warming might as well be a direct cause of human activities, such as burning coal, oil, and gas into the atmosphere (Strauss et al. 2012). Once the damage is done, sea level rise becomes an issue.

Impact on lands and Real Estates

"The projection uses Key West tidal data from 1913-1999 as the foundation of the calculation and references the year 2010 as the starting date of the projection. Two key planning horizons are highlighted: 2030 when SLR is projected to be 3-7 inches and 2060 when SLR is projected to be 9-24 inches. Sea level is projected to rise one foot from the 2010 level between 2040 and 2070, but a two foot rise is possible by 2060" (USACE 2009). The annual projected "sea level" rise for the city of Key West, for instance, describes that the city is on a path of being completely wiped out of the map.

It is reported that "one-foot rise affects only about 65,000 homes and about $37 billion in property value, a sea-level increase of three feet would put 300,000 Florida homes—around $156 billion in property value—at risk." (Guilford 2013) This statement above intriguingly depicts the number of homes that will be affected for different levels of sea-level rise. The potential losses will undeniably be catastrophic, representing billions of US Dollars.

Is annihilation imminent?

Therefore, will Key West or similar coastal Floridian cities soon go through all of the above? Will these cities submerge under the sea?

To answer such questions, let's make use of Statistics, the science of collecting, organizing, and interpreting data. We examine four coastal Floridian cities based on their last 100 years recorded Sea-Level measurements. We will try to describe sea level rise over those years using some linear expressions, and if possible foretell any future impact the lands of Florida. The selected cities are Key West, Pensacola, St-Petersburg, and Fernandina. We will compare them to see which ones hold the highest or the lowest sea level means. Maybe, we can discover which ones are actually facing the highest danger from sea level.

Additionally, we will construct linear regression models based on both the "mean" sea level and the "maximum" sea level against independent variables such as, the time in year, the average temperature, and the average rainfall for one of the four cities. The reason is that a linear equation can express the direct relationship between the independent and the response variables above in an attempt to describe the "sea level rise" for any of the four cities.

The organization of this manuscript is as follows. Data descriptions and preliminaries are presented in Chapter 2. The average Sea levels for the four are compared in Chapter 3. The time effect on seal level rise is described in Chapter 4 and 5. The fitted model for the city of Key West data are given in Chapter 6. Linear regression models and predations are discussed in chapter 7. Finally, some concluding remarks and possible future research are discussed in Chapter 8.

Data Descriptions and Preliminaries

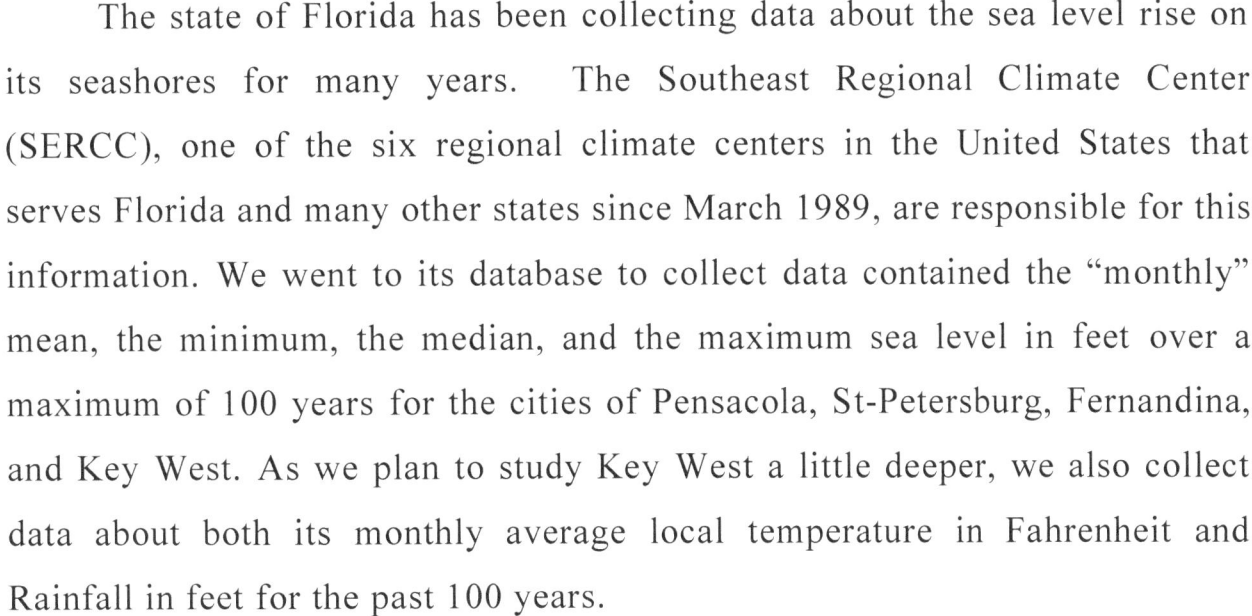

The state of Florida has been collecting data about the sea level rise on its seashores for many years. The Southeast Regional Climate Center (SERCC), one of the six regional climate centers in the United States that serves Florida and many other states since March 1989, are responsible for this information. We went to its database to collect data contained the "monthly" mean, the minimum, the median, and the maximum sea level in feet over a maximum of 100 years for the cities of Pensacola, St-Petersburg, Fernandina, and Key West. As we plan to study Key West a little deeper, we also collect data about both its monthly average local temperature in Fahrenheit and Rainfall in feet for the past 100 years.

The goal is to describe the sea level in a city. Then we may understand whether the above measurements (sea level, temperature, rainfall) increase over the years. In order to do the analysis of these data, we took the yearly average measurements, and sometime the yearly maximum measurements.

Working with the monthly data alone does not have any real statistical objective in term of years, except for yearly seasonal purposes.

The *Statistical Package for the Social Science* (SPSS) was used to analyze the data throughout this research. Analysis of Variance (ANOVA) was performed to test for significances and analyze some other essential results. Residual and mean plots were performed to describe variations or changes shown by the data. We also made use of Microsoft Excel for basic and simple analysis.

2.1 Variation between Cities:

The first objective of this research is to compare the littoral coastal cities of Key West, Pensacola, St-Petersburg, and Fernandina in terms of their average sea levels. Table 2.1 yearly organizes the data for each city in the following way:

Table 2.1: Mean seal level Descriptive Statistics

	N	Mean	Std. Deviation	95% Confidence Interval		Minimum	Maximum
				Lower Bound	Upper Bound		
Key West	95	.69090779	.238841158	.64225336	.73956222	.246000	1.185000
Fernandina	88	3.08269119	.245128447	3.03075344	3.13462895	2.363833	3.498125
St-Petersburg	68	1.10221947	.184986208	1.05744327	1.14699568	.764083	1.557500
Pensacola	92	.47425678	.219785769	.42874043	.51977312	-.027917	.914500
Total	343	1.32797555	1.077830349	1.21350571	1.44244539	-.027917	3.498125

From Table 2.1, we can see that the average sea level for Key West is .6909 ft, for Fernandina 3.0827 ft, for St-Petersburg 1.1022 ft, and Pensacola 0.4743 ft. Key West and Fernandina have data from 1914 to 2014, but contain some missing values. Pensacola has data from 1923, St-Petersburg from 1947, both with no missing information.

Histogram and box plots:

The histogram and box plots of mean sea level data for Key West, Fernandina, St-Petersburg and Pensacola are provided respectively in Figures 2.1 to 2.4. Why do we need to know these concepts? It's because some statistical conditions must be met, before we can normally make inferences.

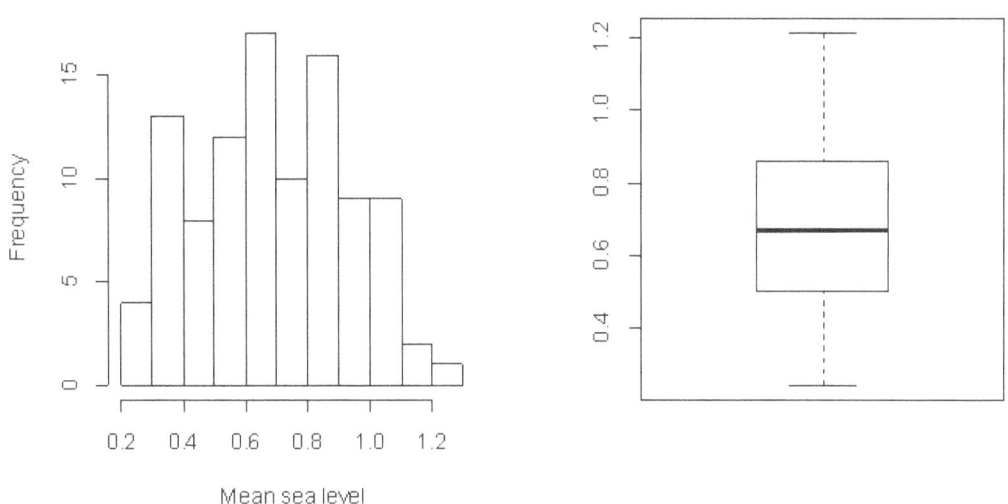

Figure 2.1: Mean sea level data in Key West

Ryan-Joiner Test for Key West:

Test statistic, Rp: 0.9902

Critical value for 0.05 significance level: 0.9871

Fail to reject normality with a 0.05 significance level.

7

Possible Outliers

Number of data values below Q1 by more than 1.5 IQR: 0

Number of data values above Q3 by more than 1.5 IQR: 0

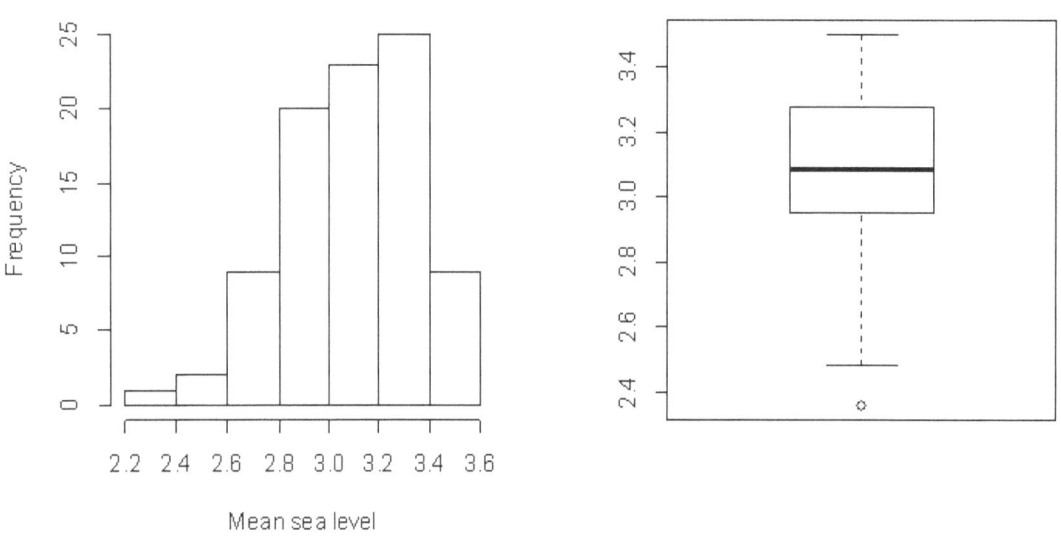

Figure 2.2: Mean sea level data in Fernandina

Ryan-Joiner Test for Fernandina

Test statistic, Rp: 0.9881

Critical value for 0.05 significance level: 0.9859

 Fail to reject normality with a 0.05 significance level.

Possible Outliers

Number of data values below Q1 by more than 1.5 IQR: 1

Number of data values above Q3 by more than 1.5 IQR: 0

8

We have one possible outlier, which is reported as the lowest sea level for the past 100 years for the city of Fernandina.

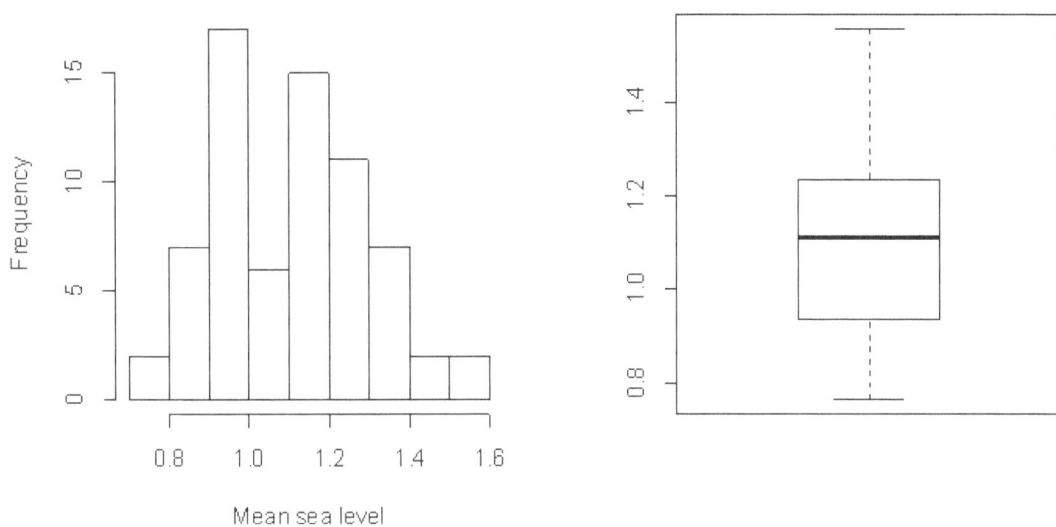

Figure 2.3: Mean sea level data in St-Petersburg

Ryan-Joiner Test for St-Petersburg

Test statistic, Rp: 0.9892

Critical value for 0.05 significance level: 0.9822

 Fail to reject normality with a 0.05 significance level.

Possible Outliers

Number of data values below Q1 by more than 1.5 IQR: 0

Number of data values above Q3 by more than 1.5 IQR: 0

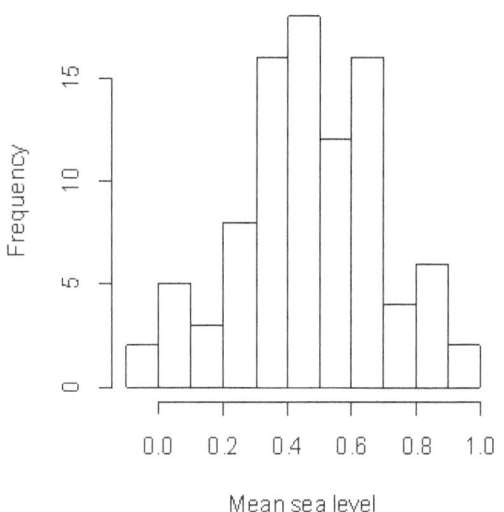
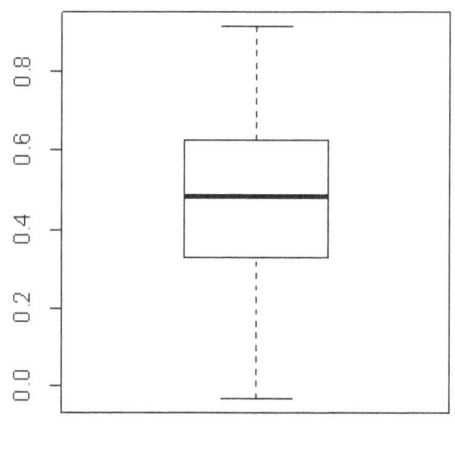

Figure 2.4: Mean sea level data in Pensacola

Ryan-Joiner Test for Pensacola

Test statistic, Rp: 0.9956

Critical value for 0.05 significance level: 0.9865

Fail to reject normality with a 0.05 significance level.

Possible Outliers

Number of data values below Q1 by more than 1.5 IQR: 0

Number of data values above Q3 by more than 1.5 IQR: 0

The boxplot of the mean sea level of Key-west, Fernandina, St-Petersburg, and Pensacola are presented respectively in the order of 1, 2, 3 and 4 in Figure 2.5. From this figure it appears that the mean sea levels are different among the four cities.

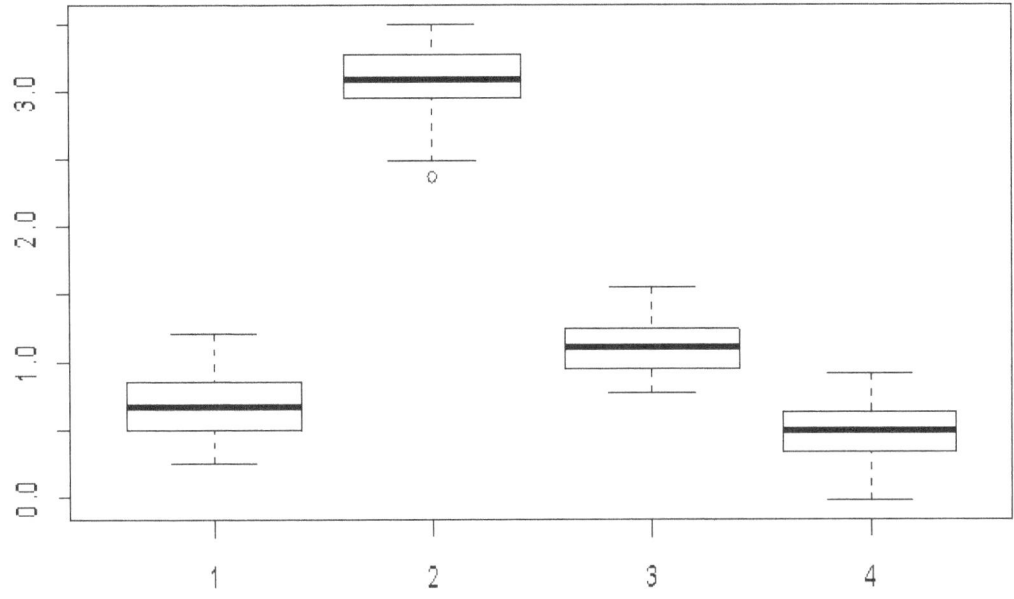

Figure 2.5: (1) Key West (2) Fernandina (3) St- Petersburg (4) Pensacola

From these figures above, it is evident that all data are roughly symmetric but the Fernandina, which is slightly left skewed. Additionally, Figure 2.5 gives an overall summary of all the data mentioned above. Therefore, all of them are normally distributed with almost zero outliers.

2.2 Data for Key West Models:

The second objective of this project is to develop some models describing the sea level rise in the city of Key West, Florida. The data are the same as described in the previous section for the city of Key West. However, to model sea level rise over time we also consider the variables year, local temperature, and the local rainfall of Key West, Florida. The years are indexed from 1 to 101, for the convenience of this study. We will fit two models: one with dependent variable "average sea level", and another with "maximum sea level".

Table 2.2 describes the yearly sea level in feet collected:

Table 2.2: Maximum sea level Descriptive Statistics

	N	Minimum	Maximum	Mean	Std. Deviation	Skewness	
	Statistic	Statistic	Statistic	Statistic	Statistic	Statistic	Std. Error
Year	101	1914	2014	1964.00	29.300	.000	.240
Mean Sea Level	95	.24600	1.18500	.6909078	.23884116	-.067	.247
Maximum Sea Level	95	.54100	1.84500	1.1404526	.29469031	-.055	.247
Mean_Temp	101	73.750	79.550	77.69191	.896818	-.720	.240
Mean_Rain	101	.0542	.1725	.111084	.0251525	.098	.240
Valid N (listwise)	95						

Table 2.2 indicates that the highest mean sea level recorded is 1.185 feet in 2013. The overall yearly mean sea level is 0.6909 feet with a standard deviation of 0.2388 feet. The average maximum sea level is 1.1405 feet with standard deviation 0.2947 feet. The overall yearly mean temperature is 77.69 degree Fahrenheit. The overall yearly mean rainfall is 0.1111 inches with standard deviation 0.0252 inches. Table 2.3 presents the Pearson's correlation coefficients among the variables.

Table 2.3: Mean sea level Correlation coefficients

		Sea Level	Year	Mean_Temp	Mean_Rain
Pearson Correlation	Mean Sea Level	1.000	.935	.315	.148
	Year	.935	1.000	.239	.161
	Mean_Temp	.315	.239	1.000	-.140
	Mean_Rain	.148	.161	-.140	1.000
Sig. (1-tailed)	Mean Sea Level	.	.000	.001	.076
	Year	.000	.	.010	.060
	Mean_Temp	.001	.010	.	.088
	Mean_Rain	.076	.060	.088	.
Pearson Correlation	Maximum Sea Level	1.000	.837	.348	.099
	Year	.837	1.000	.239	.161
	Mean_Temp	.348	.239	1.000	-.140
	Mean_Rain	.099	.161	-.140	1.000
Sig. (1-tailed)	Maximum Sea Level	.	.000	.000	.170
	Year	.000	.	.010	.060
	Mean_Temp	.000	.010	.	.088
	Mean_Rain	.170	.060	.088	.
N	Sea Level	95	95	95	95
	Year	95	95	95	95
	Mean_Temp	95	95	95	95
	Mean_Rain	95	95	95	95

From Table 2.3, we observed strong and positive correlation between mean Sea level and year (r =0.935, p-value < 0.001). Mean Sea level is also positively correlated with temperature (r = 0.315, p-value = 0.001), and not significantly correlated to rainfall (r = 0.148, p-value= 0.076). We also observed strong and positive correlation between maximum Sea level and year (r = 0.837, p-value < 0.001). Maximum Sea level is also positively correlated with temperature (r = 0.343, p-value < 0.001), and not significantly correlated to rainfall (r = 0.099, p-value= 0.170).

2.3 R-Square Change: Mean Sea level

The R-square Change method is a preliminary approach to determine the effect of variables on mean sea level. First, we built a relationship between "mean sea level" and "year", and then we check whether the additions of extra variables would provide better explanation of the data.

Table 2.4: Mean sea level R-squared changes

Model		R	R Square	Adjusted R Square	Std. Error of the Estimate	Change Statistics					Durbin-Watson
						R Square Change	F Change	df1	df2	Sig. F Change	
dimension0	1	.935[a]	.873	.872	.08544948	.873	641.391	1	93	.000	
	2	.939[b]	.882	.880	.08282184	.009	6.995	1	92	.010	
	3	.939[c]	.883	.879	.08318235	.000	.204	1	91	.652	1.208

a. Predictors: (Constant), Year

b. Predictors: (Constant), Year, Mean_Temp

c. Predictors: (Constant), Year, Mean_Temp, Mean_Rain

d. Dependent Variable: Mean Sea Level

The "R-Square Change" in Table 2.4 provides the relationships with mean sea level:

1) The baseline relationship, "Year" has a R^2 change of 0.873, with a p-value < 0.001.

2) The addition of "Temperature" provides a R^2 change with p-value= 0.009

3) The addition of "Rainfall" provides a R^2 change with a p-value < 0.001.

The first two "R^2 change" are statistically significant, but only the second model provides *information* about the prediction of the sea level rise. Adding rainfall in the presence of year and temperature does not provide any additional information.

2.4 R-Square Change: Maximum Sea level

Table 2.5: Maximum sea level R-squared changes

Model		R	R Square	Adjusted R Square	Std. Error of the Estimate	Change Statistics					Durbin-Watson
						R Square Change	F Change	df1	df2	Sig. F Change	
dimension0	1	.837ᵃ	.701	.698	.16196983	.701	218.165	1	93	.000	
	2	.851ᵇ	.725	.719	.15633411	.023	7.826	1	92	.006	
	3	.851ᶜ	.725	.716	.15717468	.000	.019	1	91	.892	1.397

a. Predictors: (Constant), Year

b. Predictors: (Constant), Year, Mean_Temp

c. Predictors: (Constant), Year, Mean_Temp, Mean_Rain

d. Dependent Variable: Maximum Sea Level

The "R-Square Change", from the above table is evaluated to verify if adding more independent variables to the equation will improve the model.

1) "Year" provides a R^2 change of 0.701 with a p-value < 0.001.

2) The addition of "Temperature" provides a R^2 change of 0.023, with a p-value= 0.006

3) The addition of "Rainfall" provides a R^2 change with a p-value < 0.001.

At alpha 0.05, the "R^2 change" model 2 is statistically significant, and provides a better estimate of the sample.

Chapter 3

Average Sea Level Comparisons

The purpose of this section is to investigate if there is a difference in the average sea level μ_i for the four cities considered in Section II It is reported that sea level might already be rising faster than expected. It might then be of interest to see which city in Florida has the highest rate of seal level rise. While we do not quite address the question of rate of sea level rise in the comparison, nevertheless we decided to compare the four cities for the present sea level. The four coastal cities selected in this study are shown in Figure 3.1.

Key West	Pensacola	St-Petersburg	Fernandina

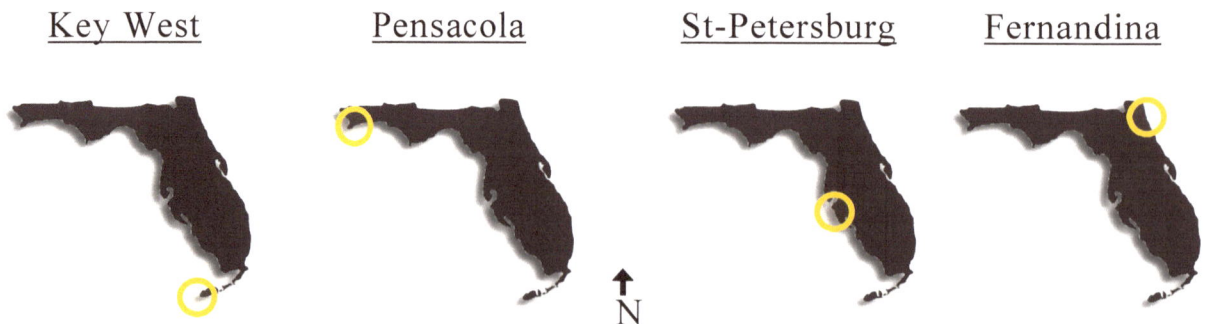

Figure 3.1: Strategic Locations of the Cities

Figure 3.1 shows the different locations of the cities mentioned above. Notice that all four coastal cities are facing sea level rise.

We have t=4 samples or cities taken independently from each other. The sample within a city is independent from the others, and vice versa. We have the following hypothesis:

Sea Level $\left|\begin{array}{l} H_0: \quad \mu_1 = \mu_2 = \mu_3 = \mu_4 = 0 \quad \text{(no difference between} \\ \text{the city means)} \\ H_1: \quad \mu_i \neq 0 \text{ for at least one } i \quad \text{(a difference \underline{exists})} \end{array}\right.$ (3.1)

3.1 The Levene Statistic

We want to compare the sea level rise for the four cities and in order to use ANOVA we need to test if the variances are equal. We can use the Levene test of Homogeneity of variances.

Hypothesis for homogeneity of Variances:

Sea Level $\left|\begin{array}{l} H_0: \quad \sigma_1^2 = \sigma_2^2 = \sigma_3^2 = \sigma_4^2 = 0 \quad \text{(equal variances)} \\ H_1: \quad \sigma_i \neq 0 \text{ for at least one } i \quad \text{(not all of the variances} \\ \text{are equal)} \end{array}\right.$ (3.2)

y_{ij}=sample observation j from city i (i=1, 2, 3, 4; and j=1, 2, 3, ..., n_i)

N = $\sum_{i=1}^{4} n_i$ the total size of all the samples

$\bar{y}_i$ = mean of sample of city i

$D_{ij}=|y_{ij} - \bar{y}_i|$ the absolute deviation of observation j from the city i

$\bar{D}_i$ = the average of the n_i absolute deviations from city i

$\bar{D}$ = average of all N absolute deviations

The Levene Statistic is then:

$$F_o = \frac{\frac{\sum_{i=1}^{4} n_i(\bar{D}_i - \bar{D})^2}{t-1}}{\frac{\sum_{i=1}^{4}\sum_{j=1}^{n_i}(D_{ij} - \bar{D}_i)^2}{N-t}}$$ 　　　　(3.3)

$$F_o = 1.972$$

Table 3.1 Test of Homogeneity of Variances

Levene Statistic	df1	df2	Sig.
1.972	3	339	.118

The Levene Statistic in Table 3.1 is 1.972 with four cities (Degree of freedom: 4-1=3) and 339 measurements (Degree of freedom: 343- 4 cities = 339). We have the *p-value* of 0.118, which supports the existence of Homogeneity of Variance. We have no strong evidence that the variances are different from each other. The data can be analyzed using ANOVA.

Since the samples between cities were independent and Levene's test for equality of variances among the cities showed no significant difference in the variances; we decided to compare the mean sea levels in the four cities using Analysis of Variance (ANOVA.)

The data on the four cities were analyzed using SPSS which gave us the following ANOVA table:

Table 3.2: Analysis of Variances of the cities

Mean Sea Level	Sum of Squares	df	Mean Square	F	Sig.
Between Groups	380.029	3	126.676	2485.367	< 0.001
Within Groups	17.278	339	.051		
Total	397.308	342			

From the ANOVA Table 3.2, we see that the p-value < 0.001. Therefore, we may conclude the mean sea levels of the four cities are different. However, we may want to know which pairs of means are different. Several methods exist to make pairwise comparisons, but we will use one of the most popular, the Scheffé's multiple comparisons:

3.2 Scheffé's Method of Multiple Comparisons:

We use the Scheffé method to do pair wise comparisons between the four cities to determine where the differences exist.

Table 3.3: Multiple comparisons of the Means

(I) City	(J) City	Mean Difference (I-J)	Std. Error	Sig.	95% Confidence Interval	
					Lower Bound	Upper Bound
Key West	Fernandina	-2.391783404*	.033402191	< 0.001	-2.48562964	-2.29793717
	St-Petersburg	-.411311685*	.035861636	< 0.001	-.51206793	-.31055544
	Pensacola	.216651013*	.033023085	< 0.001	.12386991	.30943211
Fernandina	Key West	2.391783404*	.033402191	< 0.001	2.29793717	2.48562964
	St-Petersburg	1.980471719*	.036451807	< 0.001	1.87805734	2.08288610
	Pensacola	2.608434416*	.033663057	< 0.001	2.51385526	2.70301357
St-Petersburg	Key West	.411311685*	.035861636	< 0.001	.31055544	.51206793
	Fernandina	-1.980471719*	.036451807	< 0.001	-2.08288610	-1.87805734
	Pensacola	.627962697*	.036104736	< 0.001	.52652344	.72940195
Pensacola	Key West	-.216651013*	.033023085	< 0.001	-.30943211	-.12386991
	Fernandina	-2.608434416*	.033663057	< 0.001	-2.70301357	-2.51385526
	St-Petersburg	-.627962697*	.036104736	< 0.001	-.72940195	-.52652344

*. The mean difference is significant at the 0.05 level.

From the above Table 3.3 we may conclude that all possible pairs are significantly different from each other. It appears that Fernandina has higher mean sea level measurements than the other cities with a significance level < 0.001.

A summary of the comparisons using SPSS is shown in the following table. As we can see in the table, none of the four cities have the same mean sea level with that Fernandina having the highest mean sea level (3.08 ft.) followed by St-Petersburg (1.10 ft.), Key West (0.69 ft.), and finally Pensacola with the lowest sea level (0.74 ft.).

Table 3.4: Comparison of the four cities' averages

City	N	Subset for alpha = 0.05			
		1	2	3	4
Pensacola	92	.47425678			
Key West	95		.69090779		
St-Petersburg	68			1.10221947	
Fernandina	88				3.08269119
Sig.		1.000	1.000	1.000	1.000

Means for groups in homogeneous subsets are displayed.

The following graph is summary of the Scheffe method explained above. It shows that no same cites share the same mean sea level in feet. This is called the Mean plots, which is directly built from Table 3.4.

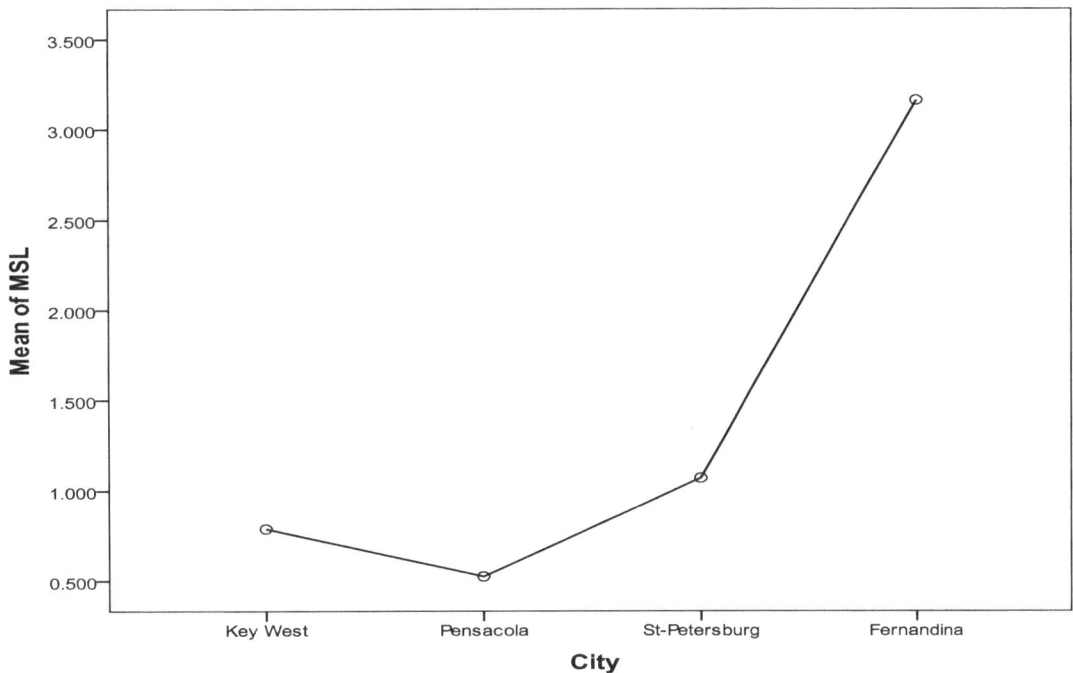

Figure 3.2: Coastal floods per city

This is a visual representation of we discussed above. Notice that Key West is higher than Pensacola, while it is lower than St-Petersburg. Also Pensacola holds the lower sea level rise in the past years. Why does Fernandina have higher sea level means than the other cities? Is it because it locates near the "Gulf Stream" directly under the influence of the North Pole, and receives more damaging effect from the melting of Greenland Ices?

Furthermore, the Global Warning Flood Risk produced by the ClimateCentral.org, describes the odds of extreme coastal floods by 2030 with sea rise from warming. We have noticed some similarities with Figure 3.4. Our study shows that Fernandina holds high sea level means, and Pensacola holds low sea level means. The ClimateCentral.org adds that the odds of extreme coastal floods by 2030 with sea rise from warming are 55% for Fernandina, and 19% for Pensacola. This is not a pure coincidence that we overly have the same conclusion.

Does Figure 3.2 imply that Fernandina will face a greater danger of sea level rise? Ideally, if all, the four cities, were identical in every way, a higher sea level rise may represent a bigger danger. However, these cities are randomly located in different locations of the coasts of Florida. Based on the given dataset, we cannot conclude whether or not that Fernandina is or will be affected by higher sea level rise than the others. As a result, this study can only conclude these four cities above differed completely in their sea level means.

Additional Approach:

Kruskal-Wallis Test of the Medians:

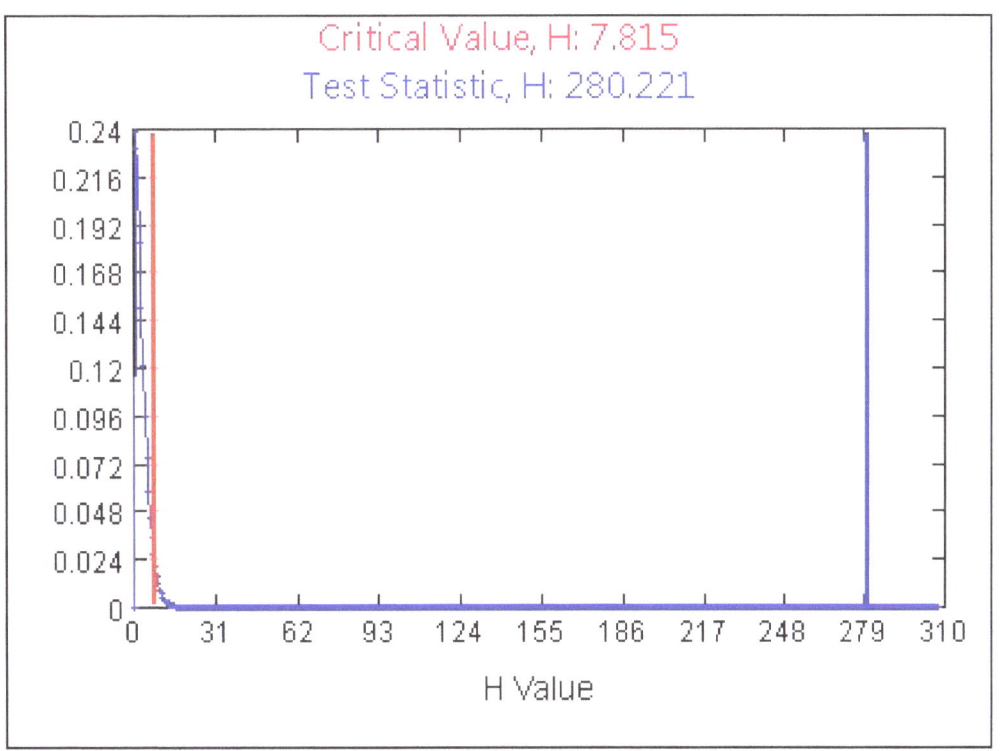

Figure 3.3: Coastal floods per city (Median)

Total Num Values: 353

Rank Sum Key West: 12808.5

Rank Sum Fernandina: 27501.0

Rank Sum St-Petersburg: 15200.0

Rank Sum Pensacola: 6971.5

Test Statistic, H: 280.2215   Critical H: 7.8147

23

P-value: 0.000

Reject equal population medians.

Data provides evidence that the samples come from populations with different medians.

Using the Maximum sea level rise of the cities:

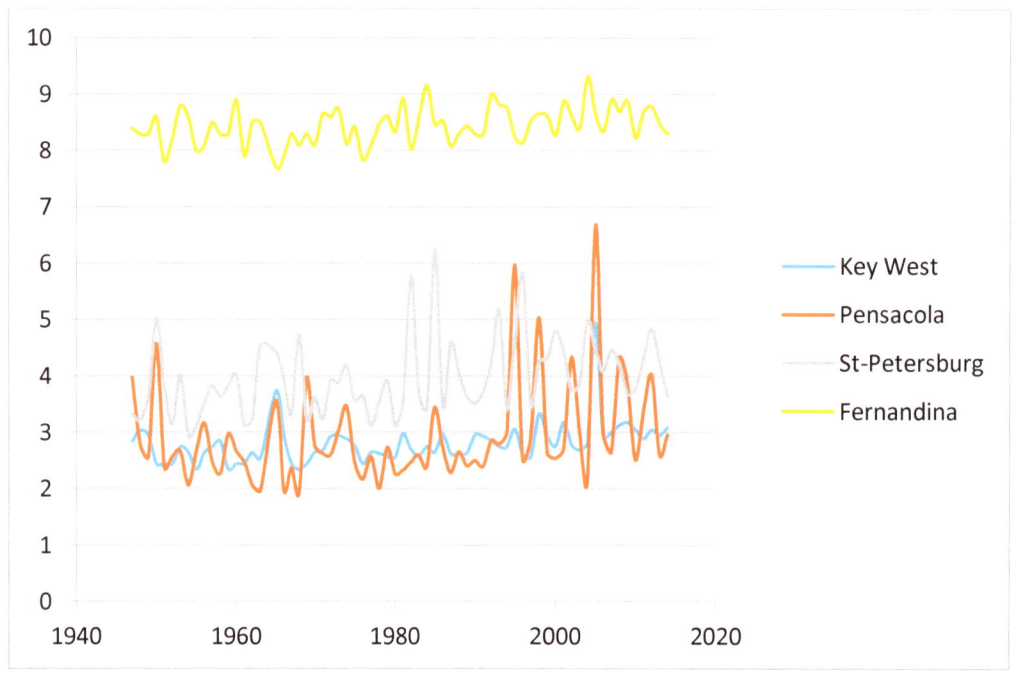

Figure 3.4: Coastal floods per city (Maximum)

Chapter 4

Key West, FL
Sea Level versus Time

The variable "year", appears to have no physical effect on the other variables in this study, but its presence is very important. Apparently, the variables sea level, temperature, and rainfall can each interact individually with "time" year or month. This interaction can be linearly increasing, decreasing, or constant which gives a sense of direction in order to make decisions on the behavior of the data. Thus in this section, we additionally have decided to examine the effect of the variable "time" on sea level rise as well as on temperature and rainfall.

Now let consider our original dataset. We want to study Key West to propose different models able to describe its sea level rise for the past 100 years. Software such as Excel allows us to generate basic linear equation models capable of doing such tasks with a 95% degree of confidence. We established the relation between seal level rise against Time order (year/month),

Temperature, and Rainfall. We examined the effect of time on sea level for the four cities, Key West, Fernandina, St. Petersburg and Pensacola. *They do not include any ice melt from Greenland or Antarctica.* The data and analysis follow:

4.1 Key West's Sea level linearly expressed as a function of time:

The data and analysis is as follow:

Future predictions of the sea level rise depend on time.

We have the following function:

S=F (T) which means "Sea-level" is a dependant function of Time

$$Y = MX + B + e \qquad\qquad (4.1)$$

where Y = <u>Sea Level</u> of a specific city, X="Year" ϵ [1, 100); M is the slope of the "straight" line; B is the intercept. The *e* stands for any error in the calculation.

As we have a data set of about 100 values, finding just one slope isn't enough to represent all the data. The slope M must be calculated in a way to represent the average of all the rises over runs between each couple in the data set. We can make use of the Least Squares Principle to determine the slope M and the intercept B.

$$M = \frac{\sum_1^n (X_i - \bar{X})(Y_i - \bar{Y})}{\sum_1^n (X_i - \bar{X})^2} \quad \text{and} \qquad (4.2)$$

$$B = \bar{Y} - M * \bar{X} \qquad (4.3)$$

Where the couple (X_i, Y_i) represents i^{th} year with a ith feet of sea level rise; $\bar{Y}$ is the average sea level rise in feet with its correspondent average $\bar{X}$ year

Now equation (4.1) becomes and represents the best fit line of the data. We also use the terminology of Regression line.

Future predictions of the sea level rise depend on time:

$$Y = \beta_0 + \beta_1 X_1 + \varepsilon \qquad (4.4)$$

where Y = Sea Level of a specific city; X_1="Year" $\in$ [1, 100); β_0 , β_1 are called the parameters (regression coefficients) of the model and need to be estimated from data. Here we assume that the errors (ε) have a normal distribution with mean 0 and constant variance σ^2.

Questions

Sea Level: **Did Sea Level increase over time?**

Proposed model: **"Sea level" against "year":**

Hypothesis:

We will test whether or not a significant and positive relationship exists between the dependant variable "sea level" and time (year).

Sea
Level

H_0: $\beta_1 \leq 0$ (Sea level is not significantly increasing)

H_1: $\beta_1 > 0$ (Sea level is significantly increasing)

If the above regression model equation is significant, it means that Sea level does increase over time. Additionally, we will be able to describe this increasing behavior of each independent toward Sea level, and ideally predict whether or not cities, like Key West, will one day slump under water, based on our dataset.

The following figure, made using Microsoft Office Excel, summarizes the yearly sea level collected in the city of Key West:

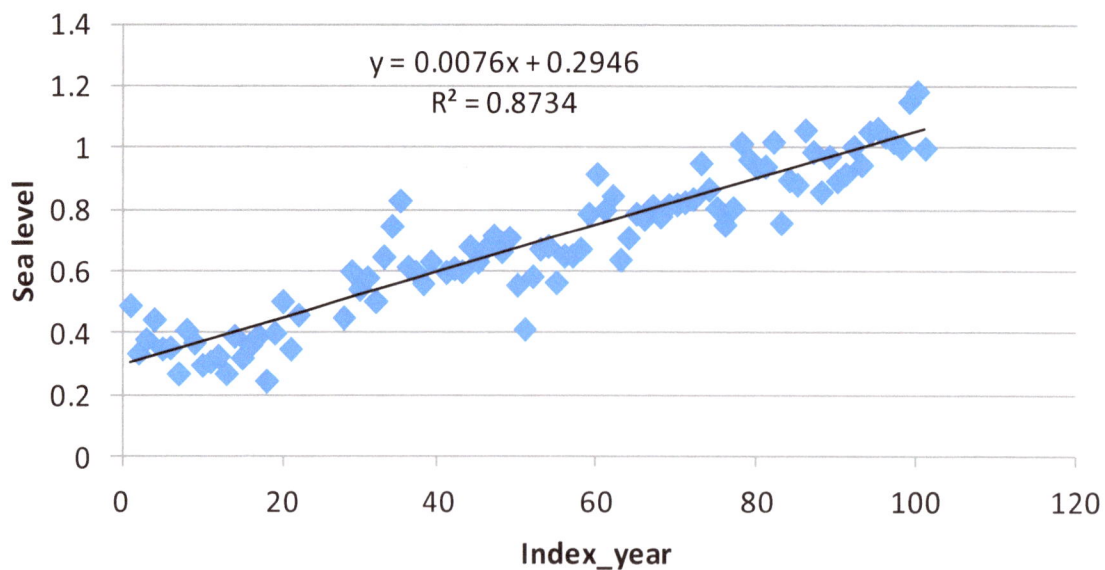

Figure 4.1: Yearly Sea level Model Plot

28

The linear relationship between sea level and year in Key West is given by:

$$y_{sea\ level} = 0.0076x_{year\_index} + 0.2946 \qquad (4.5)$$

Figure 4.1 show that sea level has a more visible increasing behavior with time. The fitted model has an R^2 of 0.873, indicating that 87.3% of the variability in sea level data is explained by the equation *(4.5)* and increases by a yearly average of 0.008 ft over the past 100 years.

The *resulting* equation *(4.5)* above <u>may</u> *only describe the sea level rise for the city of Key West in the range 1914 to 2014*, and *not be suitable to predict any future values further than 2014*. The data collected were up to 2014; we do not have data for the year 2060, 2080, or 2100 for instance. Therefore we cannot statistically predict anything for the future.

4.2 Key West's Temperature linearly expressed as a function of time:

Future predictions of the temperature of Key West depend on time:

$$Y = \beta_0 + \beta_1 X_1 + \varepsilon \qquad\qquad (4.4)$$

where Y = <u>Temperature</u> or <u>Rainfall</u> of a specific city; $X_1 =$"Year" ϵ [1, 100); β_0 , β_1 are called the parameters (regression coefficients) of the model and need to be estimated from data. Here we assume that the errors (ε) have a normal distribution with mean 0 and constant variance σ^2.

A scatter plot shows the relationship between Temperature and time is as follow:

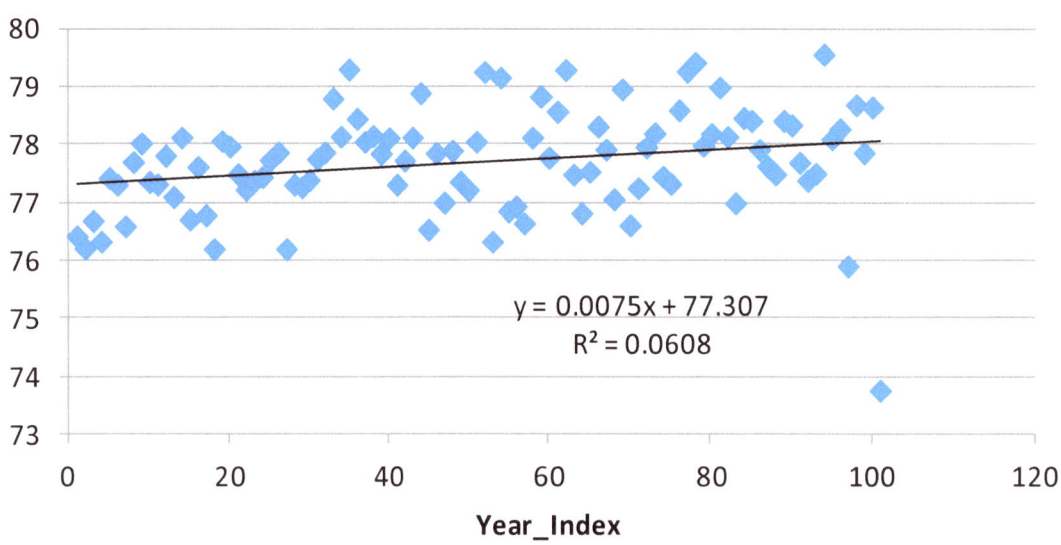

Figure 4.2: Yearly Temperature Model

That linear relationship is:

$$y_{temperature} = 0.0075 x_{year\_index} + 77.305 \qquad\qquad (4.6)$$

When rounded to three decimal places, the coefficient of x is 0.008. So for every unit increase in year, a 0.008 unit increase in mean sea temperature is predicted. However, the R^2 value for this model is quite low with only 6.1% of the variability in temperature being explained by equation (4.6).

4.3 Key West's Rainfall linearly expressed as a function of time:

A scatter plot shows the relationship between Rainfall and time is as follow:

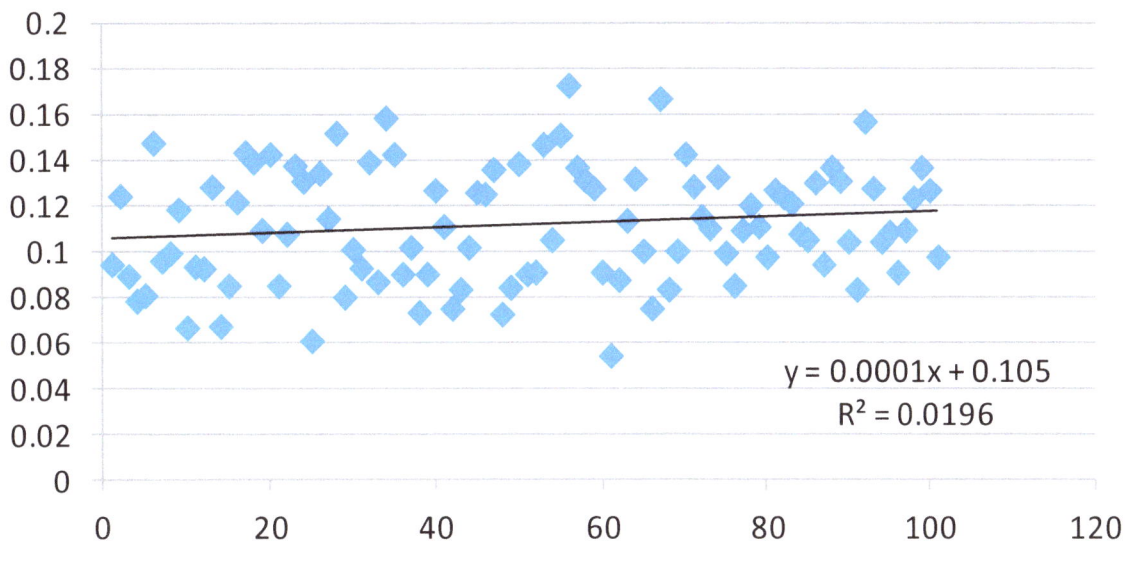

Figure 4.3: Yearly Rainfall Model

This generates the linear relationship:

$$y_{rainfall} = 0.0001x_{year\_index} + 0.105 \qquad (4.7)$$

31

Notice the coefficient of x, if round to three decimal places, is zero. This implies that rainfall and time may have no relationship. In other words, rainfall may maintain a constant value of 0.105ft each year on average.

We found no significant relationship between annual rainfall and year.

4.5 Key West's Seasonal patterns (no models):

Let's look at some seasonal patterns in the data for the city of Key West. That way, we may identify in which seasons the indices fluctuate the most.

The 100 years range is from 1914 to 2014 which contains a total of 1204 months. Notice that the number of 1118 months or sea level measurements in feet. About 93% of the total data is still available to us. Let's do a multiple comparison of months to months to determine where the differences exist, or where the sea level rise is at its most. I took the liberty to also include information about both the temperature and rainfall.

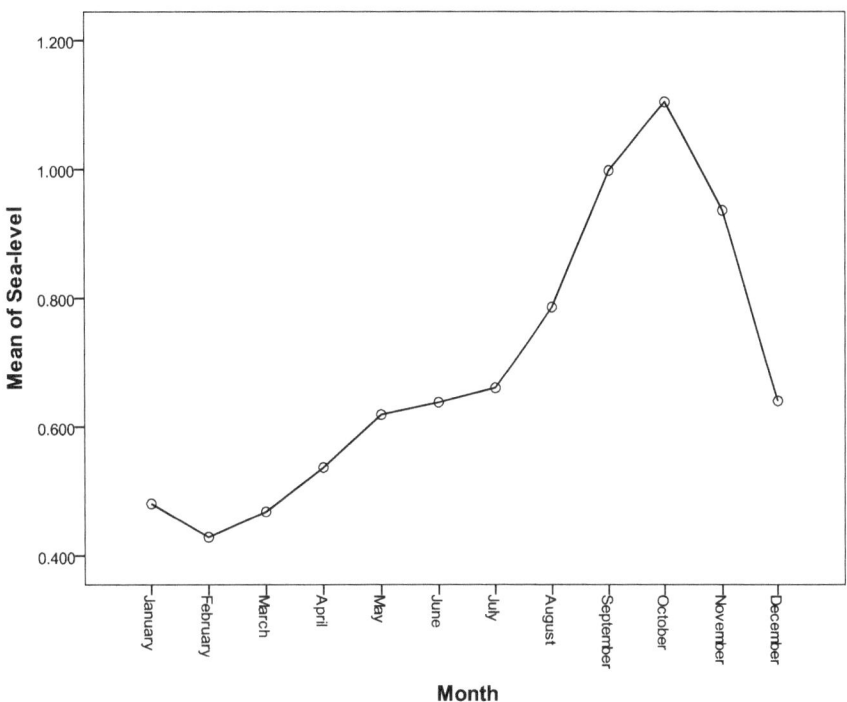

Figure 4.4: Monthly Mean Sea-level

Sea level rise is higher in the months of September, October, and November.

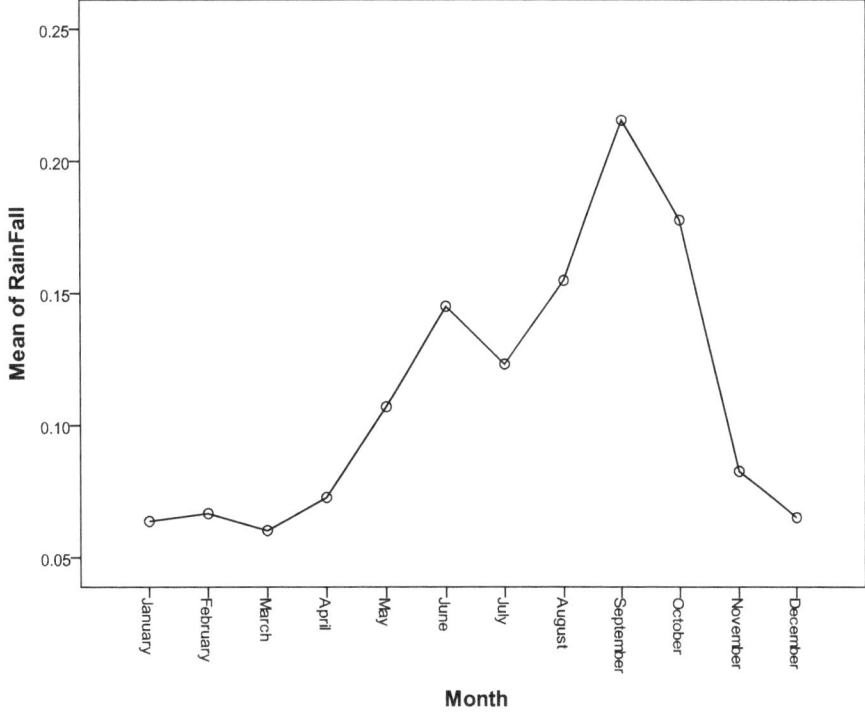

Figure 4.5: Monthly Mean Rainfall

33

Rainfall is higher in the months of August, September, and October.

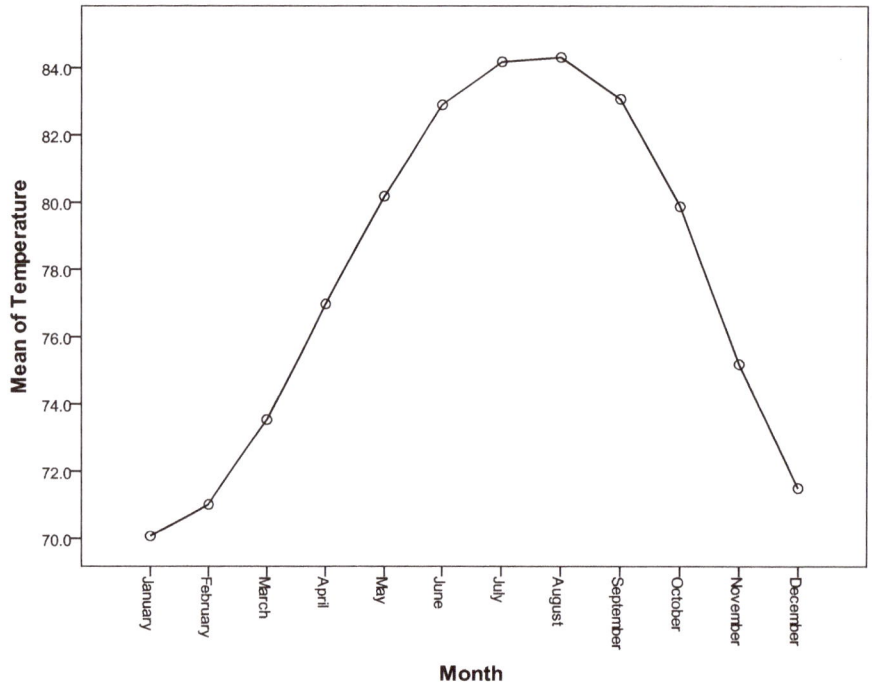

Figure 4.6: Monthly Mean Sea-level

Temperature is decreasing from September to October.

Considering the months in which sea level fluctuate the most based on our datasets:

Months	Jul-Aug	Aug-Sept	Sept-Oct	Oct-Nov
Sea Level	↗	↗	↗	↘
Rainfall	↗	↗	↘	↘
Temperature	↗	↘	↘	↘
Overall Tendency	High	Moderate-High	Moderate-Low	Low

Table 4.1: Peak Months

In a year, mean sea level rise looks nothing like temperature on average. It, additionally, also gets high as Rainfall gets high.

Note that at the beginning of the fall, the mean sea level rise is getting higher, especially in September, while Rainfall is high, and Temperature starts decreasing.

Comment

In this section, we had propose and built different models reflecting the relationship between "men sea level rise" and the independent variables, time order, temperature, and rainfall. The study was made both yearly and monthly. We did not specify any model statistics with "month"; just because the mean plots show no linear relationship when "time" is replaced by "month". We had outliers, but they did not affect the models.

Chapter 5

Coastal Cities

Sea level Rise versus Time

5.1 Fernandina's sea level

The relationship between sea level and year for Fernandina is given by:

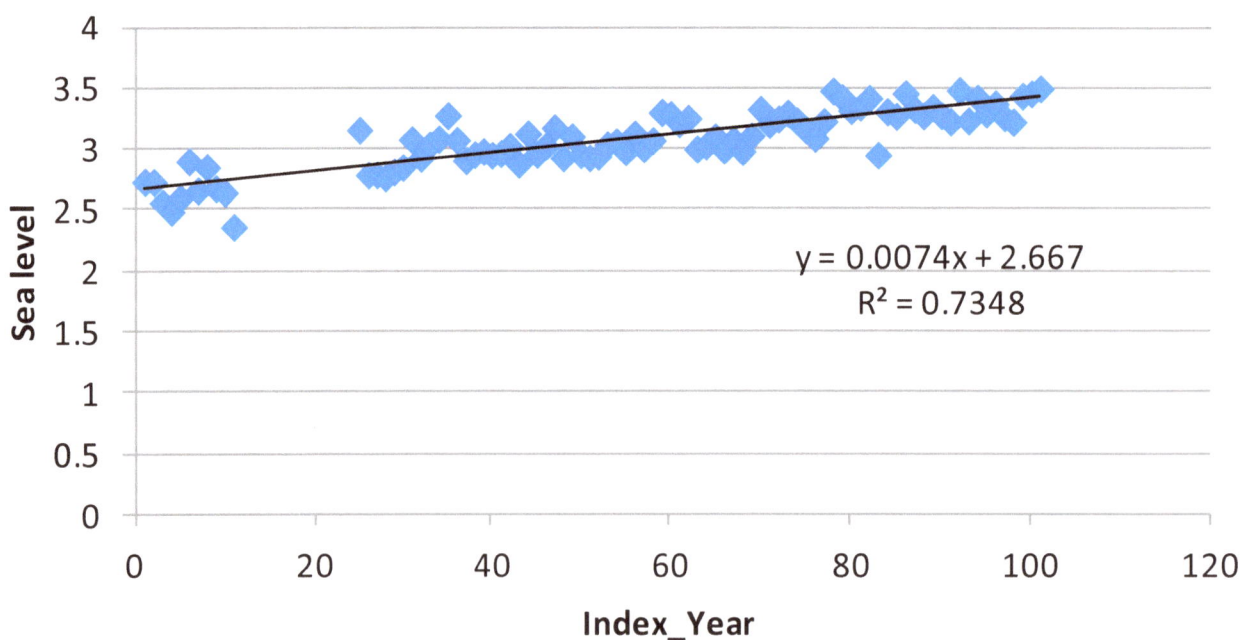

Figure 5.1: Fernandina Model Plot

The linear relationship between the two variables is estimated by:

$$y_{sea\ level} = 0.0074x_{year\_index} + 2.667 \qquad (5.1)$$

Figure 5.1 show that the sea level has an increasing behavior with time. The fitted model has an R^2 of 0.735, indicating that 73.5% of the variability in sea level data is explained by the equation *(5.1)* and increases by a yearly average of 0.0074 ft over the past 100 years.

5.2 Saint-Petersburg's sea level

The graph for St. Petersburg is given by:

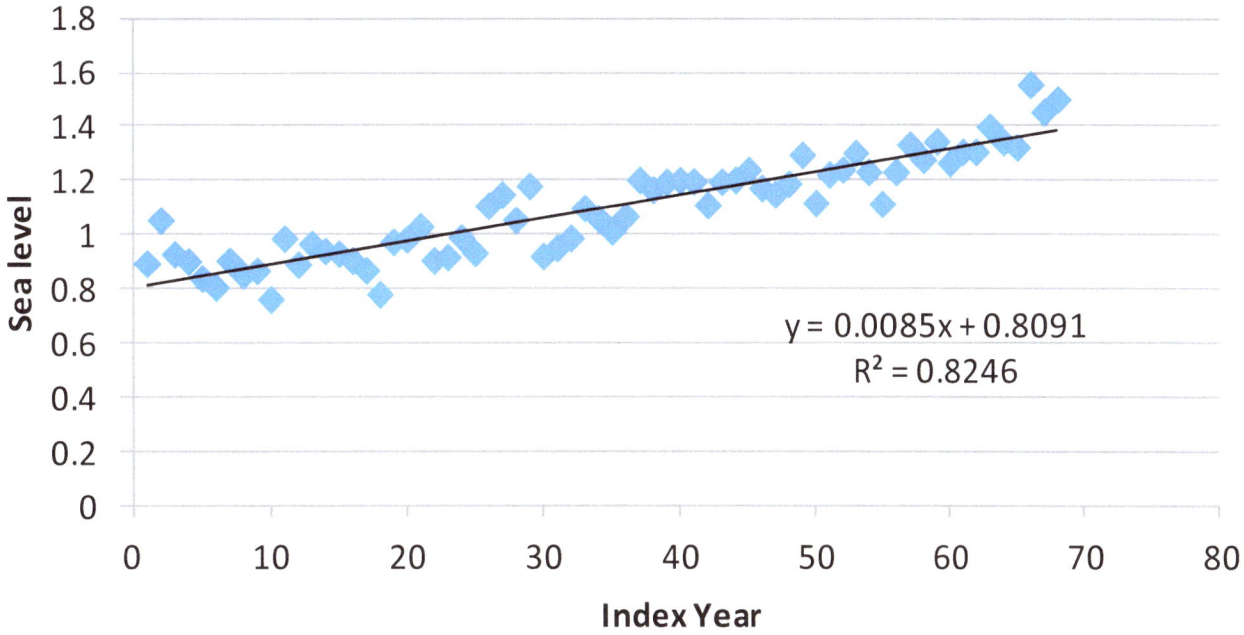

Figure 5.2: St-Petersburg Model Plot

37

The linear relationship:

$$y_{sea\ level} = 0.0085x_{year\_index} + 0.8091 \qquad (5.2)$$

Figure 5.2 show that sea level has an increasing behavior with time. The fitted model has an R^2 of 0.825, indicating that 82.5% of the variability in sea level data is explained by the equation (5.2) and increases by 0.0085 ft every year.

.

5.3 Pensacola's sea level

Finally for Pensacola, the relationship between sea level and year is given by:

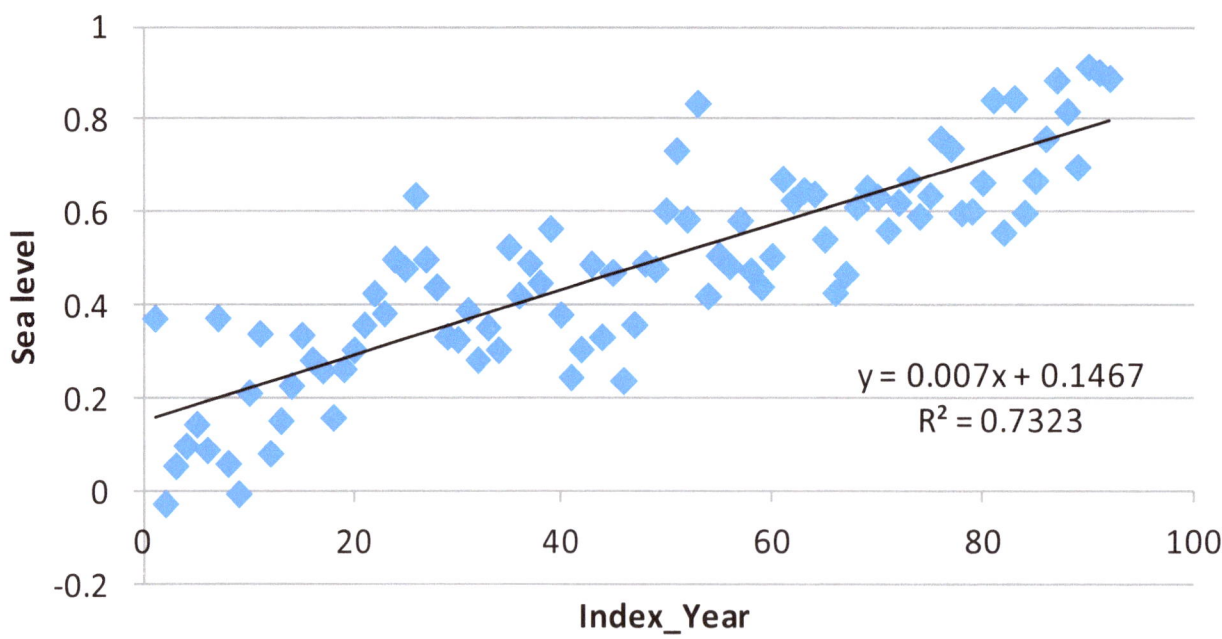

Figure 5.3: Pensacola Model Plot

The linear relationship:

$$y_{sea\ level} = 0.007x_{year\_index} + 0.1467 \qquad (5.3)$$

Figure 5.3 show that sea level has an increasing behavior with time. The fitted model has an R^2 of 0.825, indicating that 82.5% of the variability in sea level data is explained by the equation (5.3) and increases by 0.007 ft every year.

.

Summary of the Models of the cities per year:

The following figure is a visual representation of all the models against time:

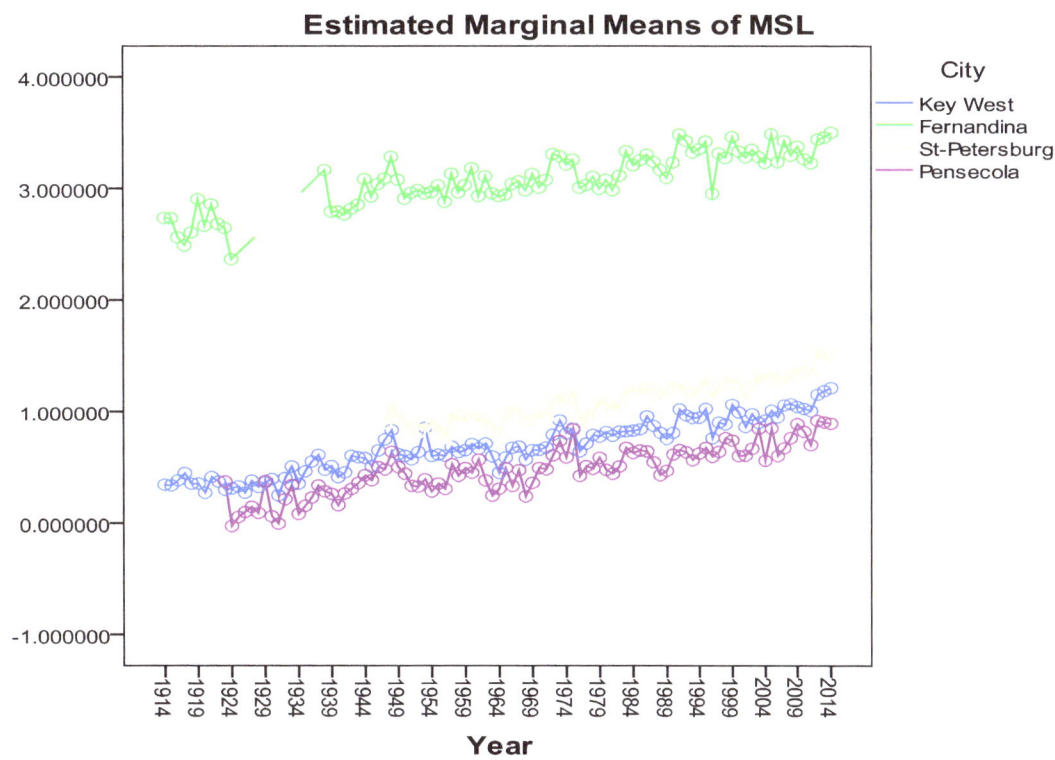

Figure 5.4: Cities' Mean Sea level Plots

Figure 5.4 gives a visual representation expanded over the years of the cities. Notice that the sea level in Key West is between Pensacola and St-Petersburg. All three of these cities have lower sea level values than Fernandina in the past years. Fernandina has higher sea levels because it starts with a higher sea level. The rise in sea level in Fernandina is actually lower than that in Pensacola and Key West based on their coefficients of the year index. The four cities differ considerably from each other.

.

Chapter 6

Key West's Sea Level Linear Regression Models

A model of equation is to express "sea level" as a function of "Year", "Temperature", "Rainfall", and "Ice Melt". Then we have the following function:

Sea Level = F (Year, Temperature, Rainfall, Ice Melt)

To avoid any bias of the beta coefficients, all "relevant" independent variables are included as much as possible in the equation, especially Time. "Recent knowledge about the lack of restraining bathymetry under Antarctica and the extent of penetration of deep fjords under the Greenland Ice Sheet means that the 4.1-6.6 feet of projected sea level rise by 2100 is most certainly low." (H. Walness) The Ice melting from the poles is due mainly to "*global warming*". We can understand that "Sea level" is rising due to so many factors. However, the presence of Ice melt from the pole has exponentially accelerated the process of sea level rise. Amazingly, Temperature is an element of "global warming". Yet, we have no need to prove here that "Temperature" and "Ice melt" might be highly correlated.

Here, we will investigate the effect of three independent variables (Year, Temperature and Rainfall) on the sea level rise for the past 100 years in Key West. The correct notation will be:

S=F (Y, T, R) which means

"Sea-level"=F (Year, Temperature, Rainfall)

The Sea level of Key West is dependant or a function of the time in year, the local temperature, and the rainfall received. The mathematical expression is the following multi-linear regression model:

$$Y = \beta_0 + \beta_1 X_1 + \beta_2 X_2 + \beta_3 X_3 + \varepsilon \qquad (6.1)$$

where Y = <u>Sea Level</u>; X_1="Year" ϵ [1, 100); X_2="Temperature"; X_3="Rainfall"; and β_0 , β_1 , β_2 , β_3 are called the parameters (regression coefficients) of the model and need to be estimated from data. Here we assume that the errors (ε) have a normal distribution with mean 0 and constant variance σ^2.

Logically, if all of the beta coefficients are negative, we should expect a decrease of the sea level measurements. Unfortunately, this idealistic "decrease" can only be achieved in the event that we, willingly, consider reducing our immense, destructive and painful, contributions to "global warming". For instance, the radical elimination of Carbone Dioxides' emissions into our lovely atmosphere... From there, we may as well make inferences for any other city similar to Key West.

Therefore, we will fit two models: one with dependent variable "average sea level", and another with "maximum sea level". The independent variables will be the "index year", "temperature", and "rainfall".

6.1 Fitting Linear Regression Model using Average Sea Level:

Using SPSS to analyze the data gave us the following results:

Table 6.1: Mean Sea level Model Selection

Model	Unstandardized Coefficients		Standardized Coefficients			95.0% Confidence Interval for B		Collinearity Statistics	
	B	Std. Error	Beta	t	Sig.	Lower Bound	Upper Bound	Tolerance	VIF
1 (Constant)	.295	.018		16.423	.000	.259	.330		
Index_Year	.008	.000	.935	25.326	.000	.007	.008	1.000	1.000
2 (Constant)	-1.683	.748		-2.250	.027	-3.168	-.198		
Index_Year	.007	.000	.911	24.743	.000	.007	.008	.943	1.060
Mean_Temp	.026	.010	.097	2.645	.010	.006	.045	.943	1.060
3 (Constant)	-1.764	.772		-2.284	.025	-3.298	-.230		
Index_Year	.007	.000	.908	24.038	.000	.007	.008	.905	1.105
Mean_Temp	.026	.010	.101	2.671	.009	.007	.046	.910	1.098
Mean_Rain	.160	.354	.017	.452	.652	-.543	.862	.941	1.063

A look at the p-values of the coefficients shows that rainfall is not significant and so we decided to drop it from consideration. The final fitted model then is:

$$\hat{Y} = -1.683 + 0.007X_1 + 0.026X_2 \qquad (6.2)$$

The number -1.683 is the constant Y intercept, the height of the regression line when it crosses the Y-axis. In other words, this is the predicted value of Sea level when the independent variable year is zero. When year is zero, Sea level is decreasing by 1.683 on average. It is significant with p-value of 0.027. The coefficient for year is .007.

So for every unit increase in year, an average of 0.007 of mean sea level units is predicted. The index Year varies from 1 to 100. It is significant with p-value of 0.000. Table 6.1 also indicates that the associated regression coefficients are not poorly estimated because of multicollinearity, or Variances Inflation Points (VIF$_j$) are less than 5.

An examination of the residuals for outliers revealed that we had one possible outlier, but removing it from the model did not change the interpretation of the model.

Table 6.2 Mean Sea level Casewise Diagnostics

Case Number	Std. Residual	Mean Sea Level	Predicted Value	Residual
51	-3.258	.41455	.6855506	-.27100059

a. Dependent Variable: Mean Sea Level

The QQ plot and residual versus fitted values are plotted in Figures 6.1 and 6.2 respectively. From these figures, we observed that both normality assumption and constant variance assumption for residuals have been met.

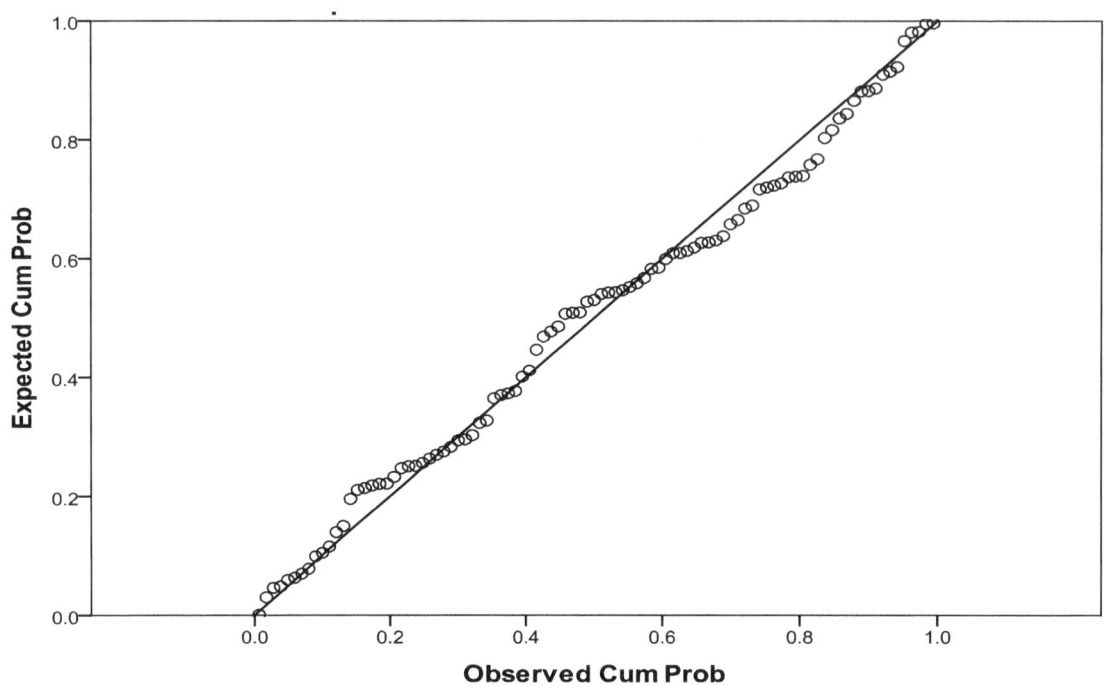

Figure 6.1: Mean Sea level Normal Probability Test

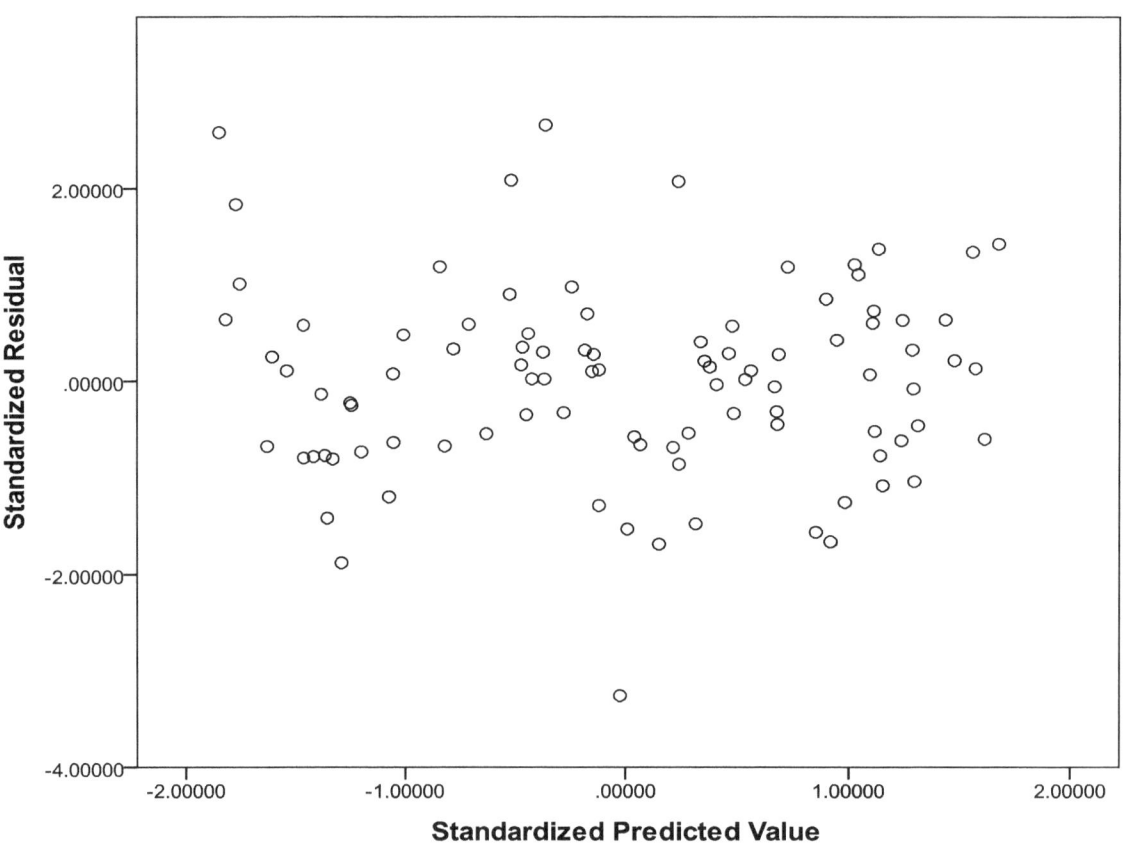

Figure 6.2: Mean Sea level Homogeneity of Variances Test

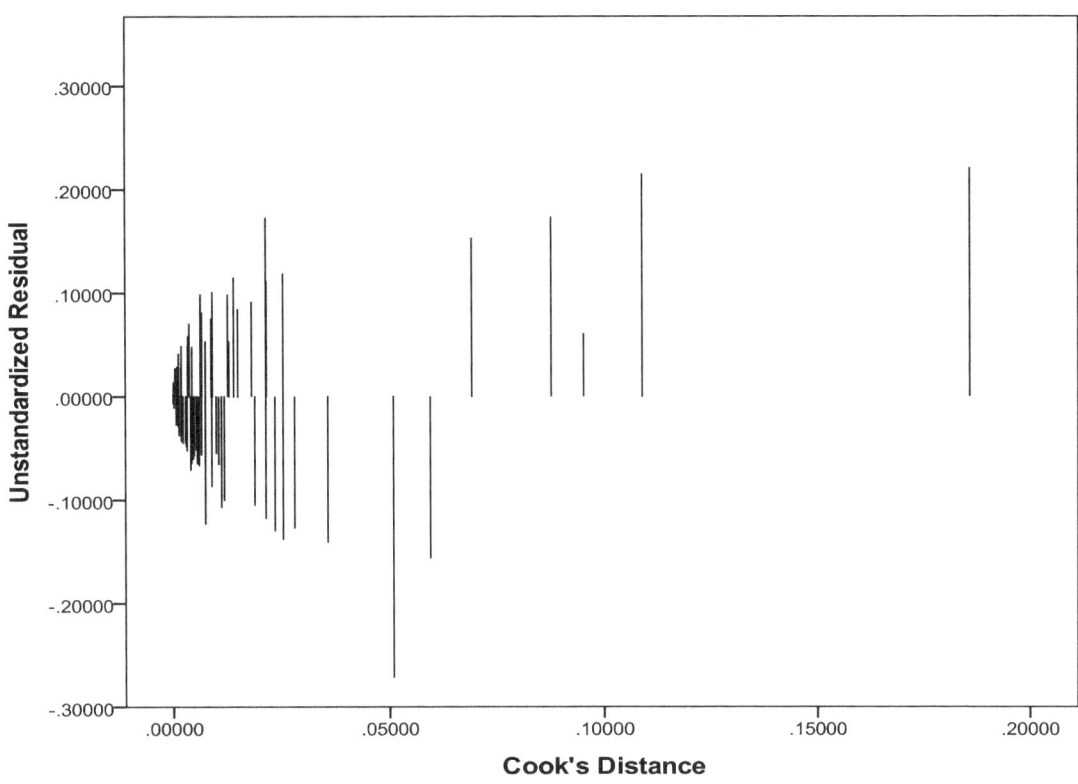

Figure 6.3: Mean Sea level Cook's Distance Influential case Test

To check the adequacy of the model, we examined Cook's Distance as shown in Figure 6.3. Note that it is in the range of $\pm$ 0.300. *Therefore, no case is influencing the model.* We have a sample size of 95; the lowest standard residual is -3.258 which correspond to case № 51.

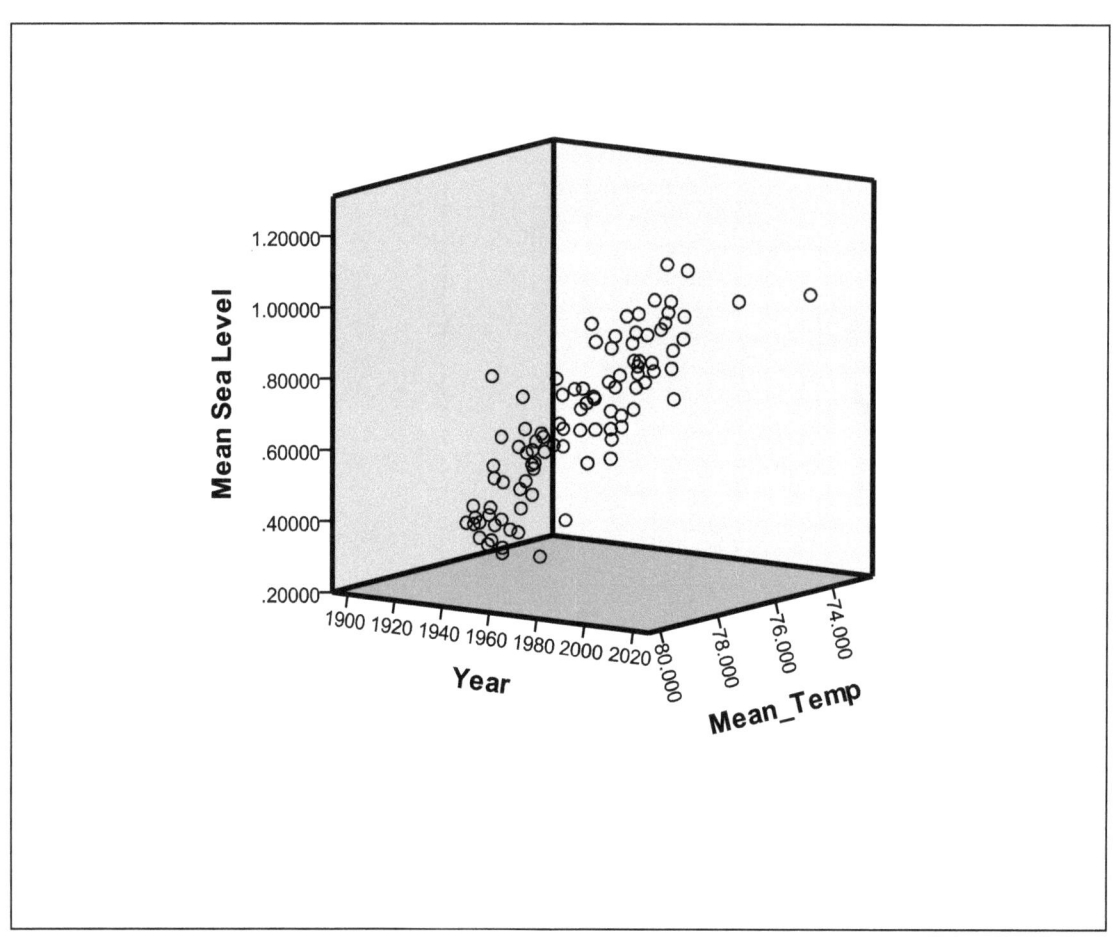

Figure 6.4: Mean Sea level in 3D

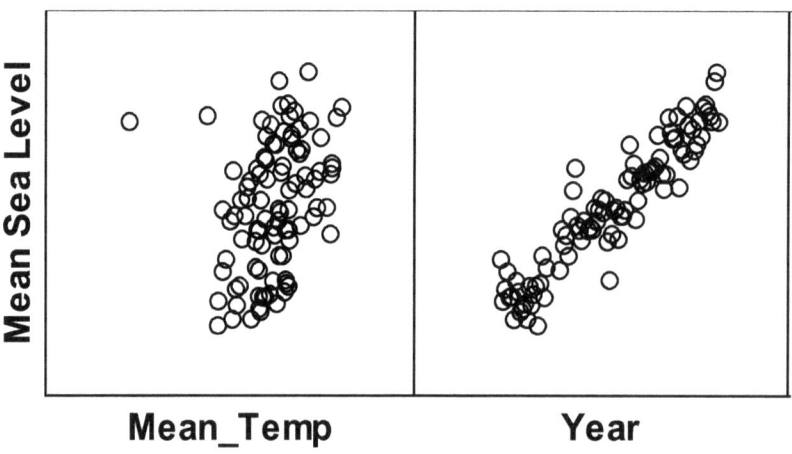

Figure 6.5: Mean Sea level in cross section

6.2 Fitting Linear Regression Model using Maximum Sea Level:

Here we decided to investigate if there is a linear relationship between maximum annual sea level in Key West and the independent variables; year, mean temperature and mean rainfall. As in the previous section, an SPSS analysis of the data showed that mean rainfall was not significant and was dropped from the model (see Table 6.3) giving us the final model as:

Table 6.3: Maximum Sea level Model Selection

Model	Unstandardized Coefficients		Standardized Coefficients			95.0% Confidence Interval for B		Collinearity Statistics	
	B	Std. Error	Beta	t	Sig.	Lower Bound	Upper Bound	Tolerance	VIF
1 (Constant)	.702	.034		20.656	.000	.635	.770		
Index	.008	.001	.837	14.770	.000	.007	.009	1.000	1.000
2 (Constant)	-3.246	1.412		-2.299	.024	-6.050	-.442		
Index_Year	.008	.001	.800	14.193	.000	.007	.009	.943	1.060
Mean_Temp	.051	.018	.158	2.798	.006	.015	.087	.943	1.060
3 (Constant)	-3.200	1.459		-2.193	.031	-6.098	-.302		
Index	.008	.001	.801	13.854	.000	.007	.009	.905	1.105
Mean_Temp	.051	.019	.156	2.709	.008	.013	.088	.910	1.098
Mean_Rain	-.091	.668	-.008	-.136	.892	-1.418	1.236	.941	1.063

$$Y = -3.246 + 0.008*X_1 + 0.051X_2 \qquad (6.3)$$

Y = $\underline{Sea\ Level}$; X_1= "Index_Year" ϵ [1, ∞); X_2= "Temperature";

As one can see both year and temperature are significant with p-value < 0.001.

49

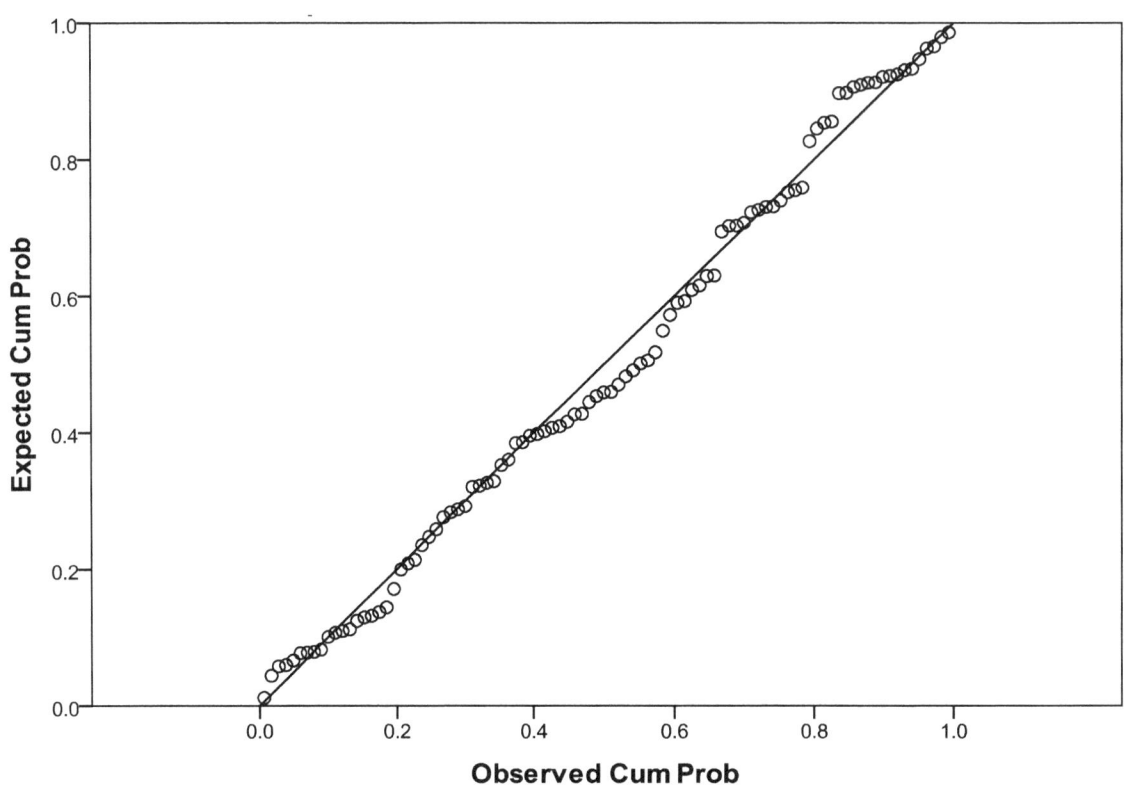

Figure 6.6: Maximum sea level Normal Probability Test

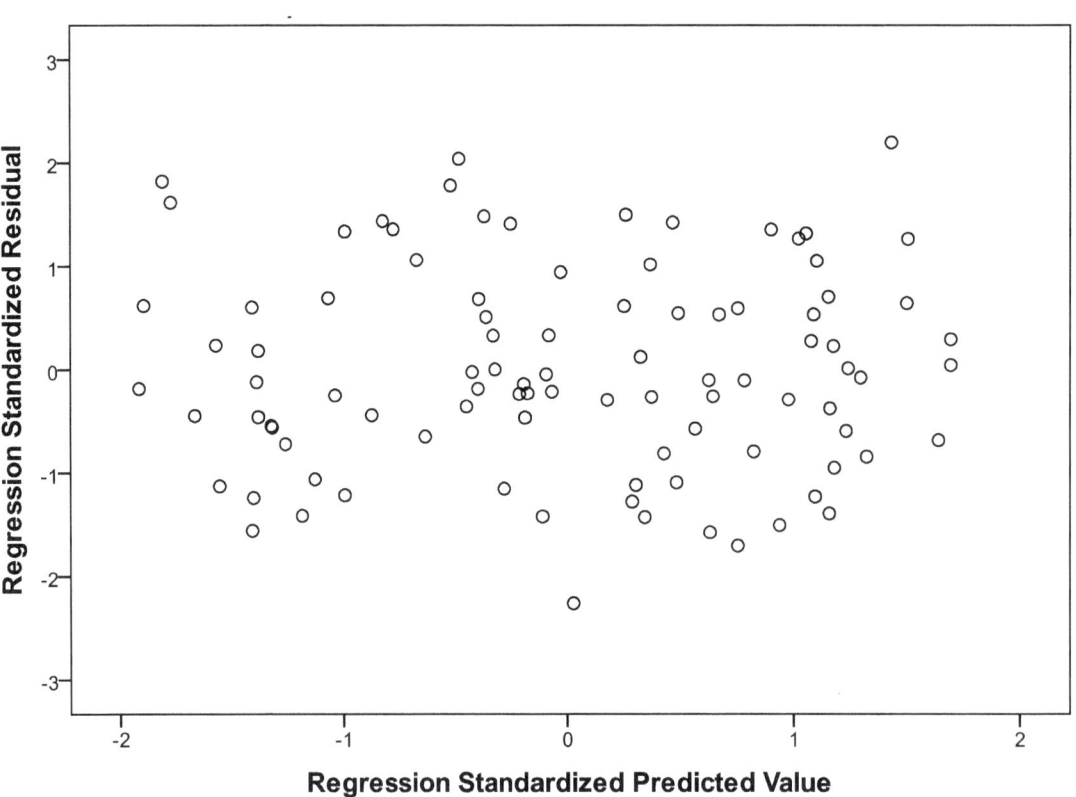

Figure 6.7: Maximum Sea level Homogeneity of Variances Test

Figures 6.6 and Figure 6.7 respectively show that the normality and constant variances assumptions for residuals are satisfied. Additionally, Figure 6.8 presents no influential case in the model.

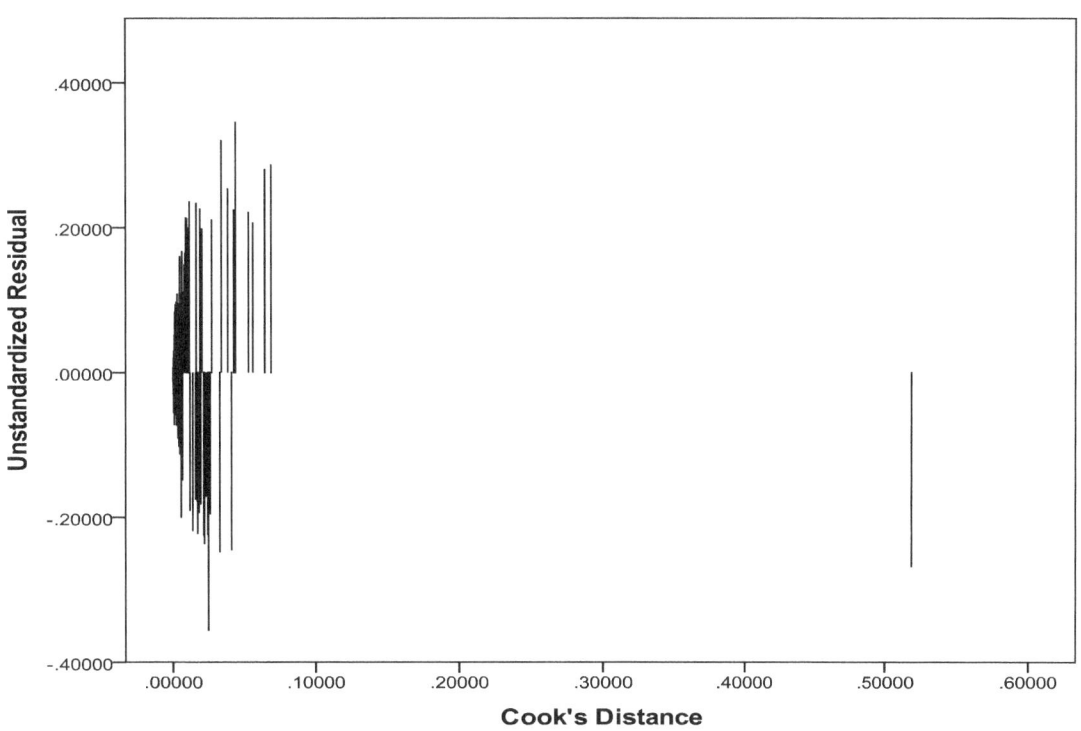

Figure 6.8: Maximum Sea level Cook's Distance Influential Test

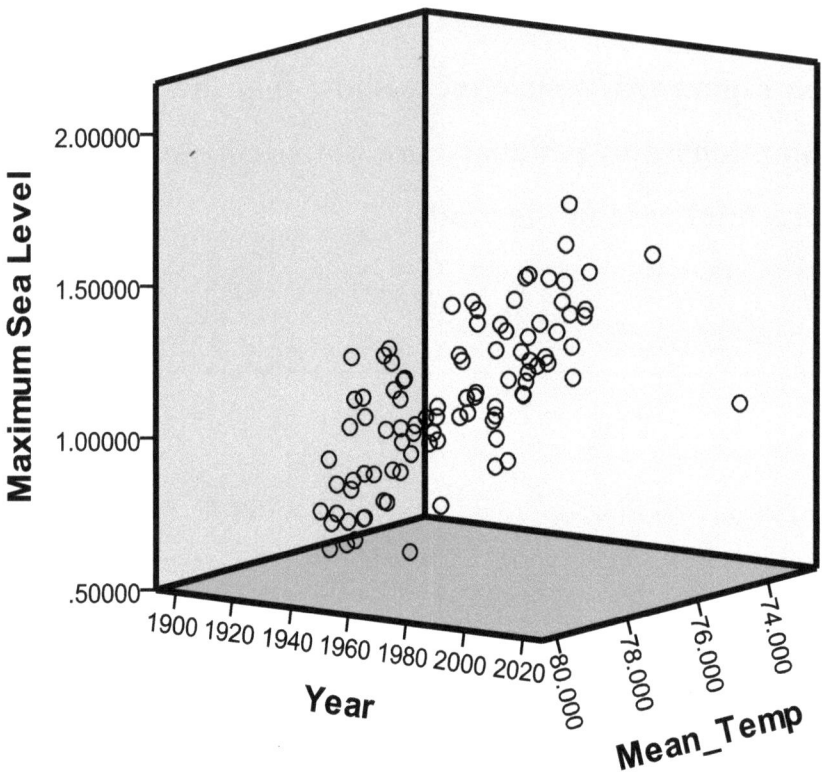

Figure 6.9: Mean Sea level in 3D

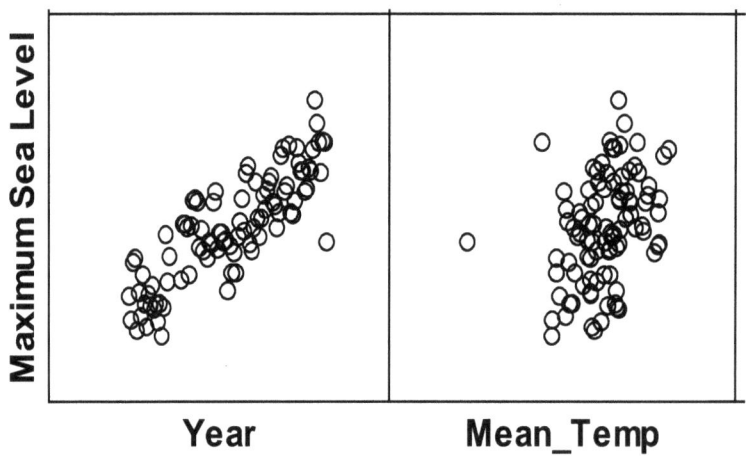

Figure 6.10: Mean Sea level in Cross Sections

Full Table of Predictions for Key West, FL

We are able to produce the following Table 6.4 to describe the sea level rise in Key West, Florida. We have the "Mean Sea Level" and the "Maximum Sea Level' models.

$$S=F (Y, T) \text{ which means that}$$

$$\text{"Sea-level"}=F (Year, Temperature)$$

Table 6.4: Sea level Model Values

YEAR	ID	Temp	Rainfall	Sea Level	Mean Sea level Model	Sea Level	Maximum Sea level Model
X1	X1	X2	X3	Y_mean	Y=-1.683+0.007*X1+0.026*X2	Ymax	Y= -3.246+0.008*X1+0.051*X2
1914	1	76.408	0.0942	0.49075	0.311	0.762	0.659
1915	2	76.2	0.1242	0.33633	0.312	0.631	0.656
1916	3	76.675	0.0892	0.38117	0.332	0.949	0.688
1917	4	76.317	0.0783	0.44558	0.329	0.972	0.678
1918	5	77.417	0.0808	0.35108	0.365	0.572	0.742
1919	6	77.3	0.1475	0.354	0.369	0.782	0.744
1920	7	76.583	0.0958	0.26925	0.357	0.651	0.716
1921	8	77.692	0.0992	0.41083	0.393	0.880	0.780
1922	9	78.017	0.1183	0.37017	0.408	0.720	0.805
1923	10	77.35	0.0667	0.29692	0.398	0.592	0.779
1924	11	77.308	0.0933	0.30858	0.404	0.772	0.785
1925	12	77.8	0.0925	0.32664	0.424	0.710	0.818
1926	13	77.092	0.1283	0.26983	0.412	0.821	0.790
1927	14	78.117	0.0673	0.39173	0.446	0.691	0.850
1928	15	76.7	0.085	0.32142	0.416	0.720	0.786
1929	16	77.608	0.1217	0.36167	0.447	0.621	0.840
1930	17	76.775	0.1433	0.39267	0.432	0.721	0.806
1931	18	76.192	0.1392	0.246	0.424	0.541	0.784
1932	19	78.058	0.1092	0.40167	0.480	0.700	0.887
1933	20	77.958	0.1425	0.50433	0.484	1.101	0.890
1934	21	77.483	0.085	0.35017	0.479	0.841	0.874
1935	22	77.217	0.1075	0.46017	0.479	0.981	0.868
1936	23	77.367	0.1375		0.490		0.884
1937	24	77.433	0.1308		0.498		0.895
1938	25	77.725	0.0608		0.513		0.918
1939	26	77.858	0.1342		0.523		0.933
1940	27	76.192	0.1142		0.487		0.856

1941	28	77.3	0.1517	0.45133	0.523	0.851	0.920
1942	29	77.258	0.08	0.60058	0.529	1.159	0.926
1943	30	77.383	0.1008	0.54355	0.539	1.159	0.941
1944	31	77.742	0.0927	0.58058	0.555	1.139	0.967
1945	32	77.858	0.1392	0.50408	0.565	0.880	0.981
1946	33	78.792	0.0867	0.64783	0.597	1.149	1.036
1947	34	78.133	0.1583	0.74767	0.586	1.290	1.011
1948	35	79.292	0.1425	0.83042	0.624	1.300	1.078
1949	36	78.442	0.09	0.61542	0.608	1.281	1.043
1950	37	78.042	0.1017	0.59883	0.605	1.031	1.030
1951	38	78.142	0.0733	0.5614	0.615	1.130	1.043
1952	39	77.833	0.09	0.63317	0.614	1.012	1.035
1953	40	78.108	0.1267		0.628		1.058
1954	41	77.3	0.1108	0.59792	0.614	0.972	1.024
1955	42	77.717	0.075	0.61042	0.632	1.061	1.054
1956	43	78.117	0.0833	0.60083	0.649	1.051	1.082
1957	44	78.883	0.1017	0.68133	0.676	1.281	1.129
1958	45	76.525	0.1258	0.63225	0.622	1.340	1.017
1959	46	77.842	0.125	0.676	0.663	1.021	1.092
1960	47	76.992	0.1358	0.71718	0.648	1.110	1.057
1961	48	77.883	0.0725	0.665	0.678	1.110	1.110
1962	49	77.35	0.0842	0.70992	0.671	1.061	1.091
1963	50	77.208	0.1383	0.5567	0.674	1.071	1.092
1964	51	78.042	0.09	0.41455	0.703	0.792	1.142
1965	52	79.25	0.0908	0.58458	0.742	1.041	1.212
1966	53	76.317	0.1467	0.6735	0.672	0.890	1.070
1967	54	79.15	0.105	0.68242	0.753	1.002	1.223
1968	55	76.842	0.1508	0.5655	0.700	0.890	1.113
1969	56	76.925	0.1725	0.65142	0.709	1.090	1.125
1970	57	76.633	0.1367	0.65208	0.708	1.172	1.118
1971	58	78.117	0.1309	0.67408	0.754	1.300	1.202
1972	59	78.825	0.1275	0.78825	0.779	1.120	1.246
1973	60	77.767	0.0908	0.91683	0.759	1.441	1.200
1974	61	78.567	0.0542	0.801	0.787	1.481	1.249
1975	62	79.283	0.0875	0.84633	0.812	1.051	1.293
1976	63	77.475	0.1133	0.63942	0.772	1.012	1.209
1977	64	76.808	0.1317	0.71033	0.762	1.139	1.183
1978	65	77.525	0.1	0.78842	0.788	1.392	1.228
1979	66	78.308	0.075	0.77258	0.815	1.192	1.276
1980	67	77.908	0.1667	0.81317	0.812	1.090	1.263
1981	68	77.05	0.0833	0.78	0.796	1.192	1.228
1982	69	78.95	0.1	0.81708	0.853	1.320	1.332
1983	70	76.6	0.1425	0.81908	0.799	1.241	1.221
1984	71	77.242	0.1283	0.82633	0.822	1.349	1.261
1985	72	77.95	0.115	0.83642	0.848	1.392	1.305
1986	73	78.192	0.11	0.95258	0.861	1.422	1.326
1987	74	77.442	0.1325	0.86825	0.848	1.281	1.296
1988	75	77.317	0.0992	0.80675	0.852	1.261	1.297
1989	76	78.592	0.085	0.75258	0.892	1.139	1.370
1990	77	79.258	0.1091	0.807	0.917	1.222	1.412
1991	78	79.408	0.12	1.01642	0.928	1.540	1.428
1992	79	77.975	0.1108	0.96342	0.897	1.579	1.363
1993	80	78.183	0.0975	0.93875	0.910	1.340	1.381

1994	81	78.983	0.1267	0.94208	0.938	1.372	1.430
1995	82	78.133	0.1236	1.02133	0.922	1.596	1.395
1996	83	76.983	0.1208	0.75892	0.900	1.222	1.344
1997	84	78.458	0.1075	0.89842	0.945	1.212	1.427
1998	85	78.408	0.105	0.88308	0.951	1.287	1.433
1999	86	77.908	0.13	1.05917	0.945	1.582	1.415
2000	87	77.617	0.0942	0.98933	0.944	1.497	1.408
2001	88	77.483	0.1367	0.85958	0.948	1.454	1.410
2002	89	78.408	0.1308	0.974	0.979	1.454	1.465
2003	90	78.333	0.1042	0.89525	0.984	1.340	1.469
2004	91	77.683	0.0833	0.91742	0.974	1.356	1.444
2005	92	77.375	0.1567	1.00633	0.973	1.471	1.436
2006	93	77.492	0.1275	0.947	0.983	1.454	1.450
2007	94	79.55	0.1042	1.054	1.043	1.573	1.563
2008	95	78.092	0.1083	1.06567	1.012	1.845	1.497
2009	96	78.267	0.0908	1.0395	1.024	1.717	1.514
2010	97	75.9	0.1092	1.02225	0.969	1.612	1.401
2011	98	78.683	0.1233	1.00292	1.049	1.445	1.551
2012	99	77.85	0.1367	1.1515	1.034	1.618	1.516
2013	100	78.642	0.1267	1.185	1.062	1.612	1.565
2014	101	73.75	0.0975	1.001	0.942	1.061	1.323

Chapter 7

Key West's Sea Level
Possible Forecasts

Table A contains the sea level values calculated alongside the observed values of the variables. These values were obtained by either using mean sea levels or maximum sea levels against the independent variables such as time, temperature, and rainfall. Rainfall was dropped from the equation. Therefore, it compiles only values from 1914 until 2014 depending on two independent variables "time or year" and "temperature". Any predictions beyond the year 2014 based on the data, would be unreliable and fictive. We can't predict the future, because of so many biases and lack of data from the future.

Additionally, temperature and sea level rise might be affected by some other factors unseen by our regression models of the data. Their relations with the independent variable time (or year) might not even be linear. As for the variable "year", it is constantly increasing with the same rate *no matter what*.

7.1 Predicting the future

Assuming the all indices above follow linear model equations, we won't have any problem estimating with a certain confidence the future values. As we do not know the future temperature measurements, we may use the equation *(4.6)* from chapter 4, to predict some future local temperature values for the city of Key West.

Checking first for significance:

Table 7.1a Mean Temperature ANOVA

Model	Sum of Squares	df	Mean Square	F	Sig.
1 Regression	4.890	1	4.890	6.409	.013[a]
Residual	75.538	99	.763		
Total	80.428	100			

a. Predictors: (Constant), Index

b. Dependent Variable: Mean_Temp

Table 7.1b Mean Temperature Model Coefficients [a]

Model	Unstandardized Coefficients		Standardized Coefficients		
	B	Std. Error	Beta	t	Sig.
1 (Constant)	77.307	.175		441.418	.000
Index	.008	.003	.247	2.532	.013

a. Dependent Variable: Mean_Temp

The ANOVA Table 7.1a shows that, at a 0.05 level of significance, the regression equation *(4.5)* is significant. At that same level of significance, we may not drop any of its coefficients from Table 7.1b. Thus, using equation (4.6), we can predict temperatures for the years between 2015 and over, represented in the third column of Table 7.2. The last two columns of Table 7.2 gave the predicted mean and maximum Sea levels for Key West.

7.1 Tentative Table to predict future values

Table 7.2: Suggestive Prediction Table

YEAR	Index	Temperature	Mean Sea level	Maximum Sea level
-	X1	X2 = 0.0075X1 + 77.307	Y=-1.683+0.007*X1+0.026*X2	Y= -3.246+0.008*X1+0.051*X2
2015	102	78.072	1.061	1.554
2020	107	78.1095	1.097	1.596
2025	112	78.147	1.133	1.638
2030	117	78.1845	1.169	1.680
2035	122	78.222	1.205	1.722
2040	127	78.2595	1.241	1.764
2045	132	78.297	1.277	1.806
2050	137	78.3345	1.313	1.848
2055	142	78.372	1.349	1.890
2060	147	78.4095	1.385	1.932
2065	152	78.447	1.421	1.974
2070	157	78.4845	1.457	2.016
2075	162	78.522	1.493	2.058
2080	167	78.5595	1.529	2.100
2085	172	78.597	1.565	2.142
2090	177	78.6345	1.601	2.184
2095	182	78.672	1.637	2.226
2100	187	78.7095	1.673	2.268
2105	192	78.747	1.709	2.310

Table 7.2 is an extract of the Data Values of Key West in the Appendix of page 43. The appendix tabulates the results obtained by this research about the city of Key West, Florida. It shows that the average mean sea level rise for Key West will be 1.097 ft. with maximum seal level of 1.596 ft. under a temperature of 78.1F^{o} for the year 2020. In 2060, that number will become 1.385 ft. on average with a maximum of 1.932 ft. under a temperature of 78.4F^{o} with respect to the range predicted by scientists of a minimum 1.25 feet to a maximum 2.25 feet. Regrettably for Florida, those digits become alarming for the year 2100: average sea level rise 1.673 ft., maximum sea level rise

2.268 feet. The range predicted by scientists on Figure 222 is even worse with a minimum 3.00 feet to a maximum 6.00 feet. Overly, our Table of predictions predicts as well high values. Our data was based on the city of Key West, Florida.

7.2 Figure presenting the models of Prediction:

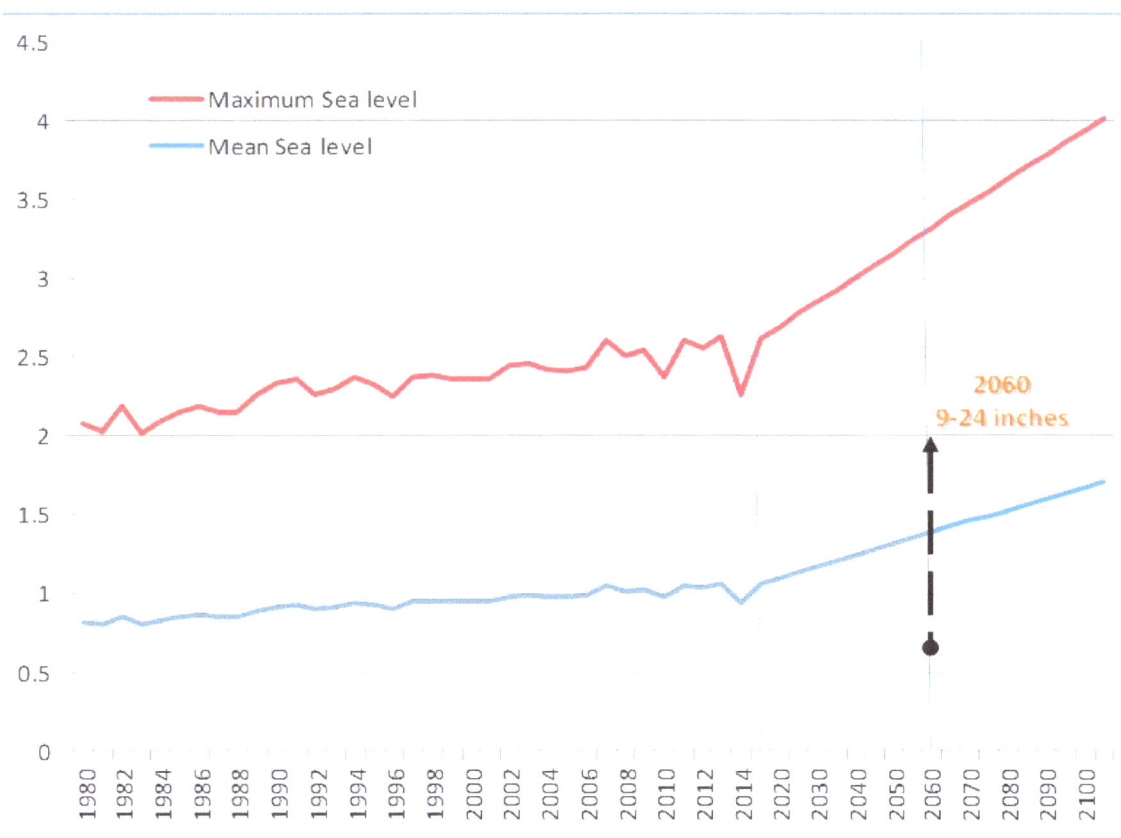

Figure 7.1: Prediction Comparison

The Figure 7.1 above gives a visual description of the appendix. Apparently, until 2014, we have a constant increase of the average sea level rise. Especially after 2015, the increase tremendously accelerated. The average mean sea level rise found in this study falls in the same range predicted by scientists.

We were able to build a linear model based on the past 100 years of measurements for the city of Key West, but our model can't predict values beyond 2014. Some outside sources predict the sea level to be in the range of 0.50 ft to 0.92 ft for the year 2030, from 1.25 ft to 2.25 ft for 2060 (a minimum of 150% increase from 2030), and from 3 ft to 6 ft for 2100 (a minimum of 500% increase from 2030, and a minimum of 140% increase from 2060). These values do not indicate linear trends between 2030 and 2100. The exact rate at which sea level is increasing might even be unpredictable, but we just can have ideas about the ranges until 2100. After 2100, the values for sea level will go sky high.

With the permission of Ben Strauss from Climate Central.org, I was able to produce the following figure to express the above results.

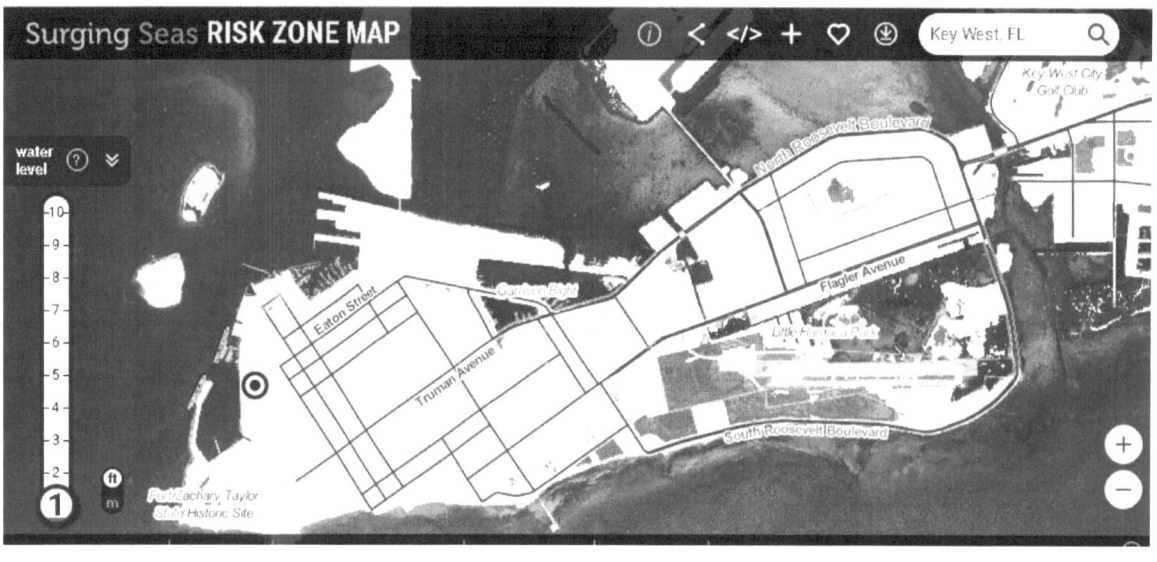

Figure 7.2a: Key West at 1 foot of Sea level rise

Figure 7.2b: Key West at 4 feet of Sea level rise

Figure 7.2c: Key West at 6 feet of Sea level rise

We have no words to explain these sad images about the future of South Florida if the event that sea level is incessantly rising up. Once the Sea Level reaches 10 (ten) feet, the state is out of the map. This is not a good destiny for the sunny and beautiful state of Florida.

Chapter 8

South Florida Cities Summary and Conclusion

Overall, the sample sizes of the data collected for Key West, Pensacola, St-Petersburg, and Fernandina were reasonable to make inferences. We had *few missing* values, but they were no greater than 15% of the data. We had outliers, but they did not affect the models.

The average sea level in Key West, Pensacola, St-Petersburg, and Fernandina differ considerably from each other. Pensacola has the lowest sea level. We considered the "average" sea level and the "maximum" sea level, which produced two-different linear regression models with independent variables year and temperature. We have made attempt to fit regression models for sea level rise data in Key West. However, following similar procedures that described in Section 4, one can fit regression models for sea level in Pensacola, St-Petersburg, and Fernandina. What will happen to Key West? Or what will happen to Florida coasts?

Overlay, this study was focused on finding the linear relationships between variables, and the comparisons between some selected cities. The conclusions about the future of sea levels in Florida are restricted to the *data sets* considered and *models* developed in this paper. To make any definite statement, one might need more data and need to fit different kind of models.

All of the models in this paper, especially the ones in Section 4, were fitted with the assumption the error terms of the models are independently and identically distributed (iid). However, since the above observations are time dependent, these models can further be predicted using the *time series models* which are part of our future research.

Some other Sea level Regression Models with time:

Dr. Harold Wanless, from the University of Miami, is one of the well known leading scientists dedicated to the study of sea level rise in Florida. He explained that sea level is rising faster than expected in South Florida and will rise about a foot each decade after 2100. With his permission, the next page develops his ideas on the` matter followed by a map from Ben Strauss.

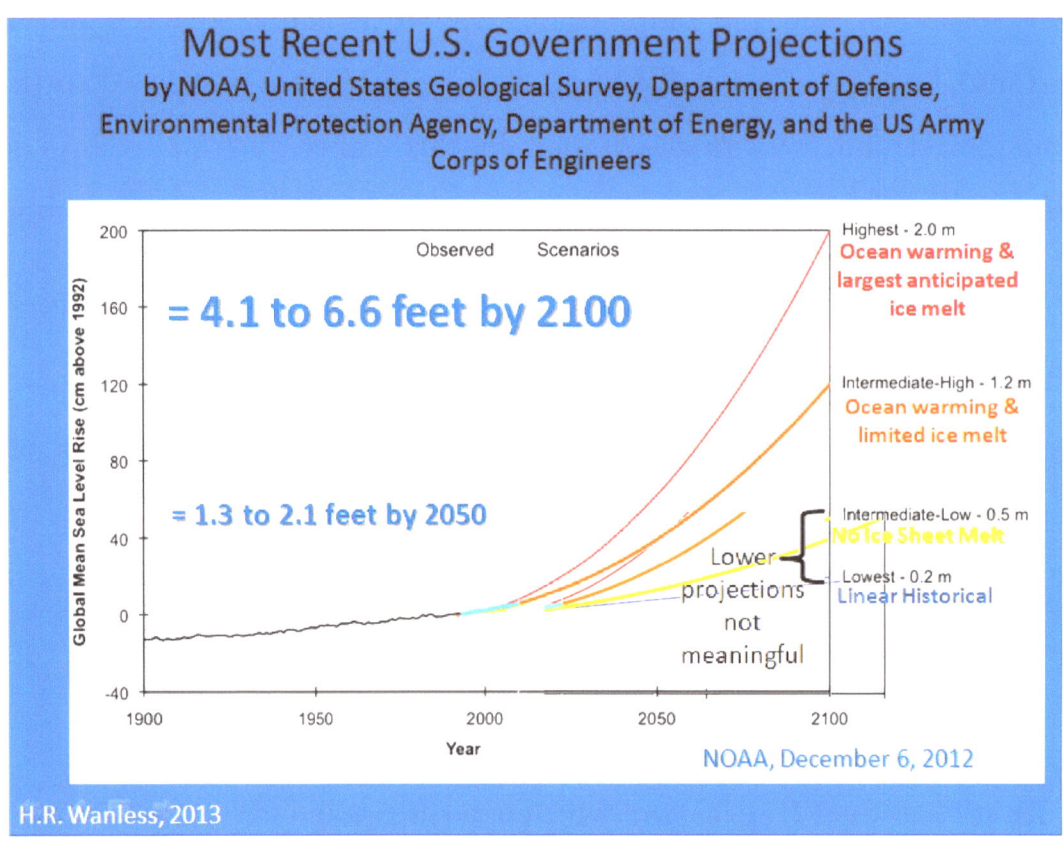

Figure 8.1: Multiple Sea level Scenarios for South Florida

Here's...a graph from the most recent US Government sea level projections (the lower two are meaningless as they do not include any ice melt from Greenland or Antarctica). Recent knowledge about the lack of restraining bathymetry under Antarctica and the extent of penetration of deep fjords under the Greenland Ice Sheet means that the 4.1-6.6 feet of projected sea level rise by 2100 is most certainly low.

Cheers,

Hal Wanless

In September 2010, scientists from University of Miami, FIU, FAU, the Corps of Army Engineers, and others met to come to a consensus on the most probable minimum and maximum sea level rise that should be used for the coming century for planning in the southeast Florida Counties, Monroe, Miami-Dade, Broward, and Palm Beach. They reached unanimous consensus on the following:

• Minimum and maximum sea level rise that should be used for planning purposes are:
  • by year 2030 – 6" to 11" (0.5 to 0.92')   [15 to 28 cm]
  • by year 2060 – 15" to 27" (1.25' to 2.25') [38 to 68 cm]
  • by year 2100 – 36" to 72" (3 to 6')   [30 to 60 cm]

• Sea level rise will continue to accelerate through this century and well into the next. If sea level has risen 3 feet by 2100, it will be rising at 0.6 feet (7.2") per decade at that point. If sea level has risen six feet by 2100, it will be rising at 1.2 feet (14.4") per decade at that point – and accelerating!

•It is very unlikely that sea level rise will be less than the minimum projections

• It is very possible that sea level rise will be greater than the maximum projections depending on (a) whether or not humanity quickly begins reducing greenhouse gas input, (b) further rates of acceleration of ice melt and associated reinforcing feedbacks, (c) timing and extent of greenhouse gas input from collapsing tundra and permafrost, (d) slowing of the Florida Current/Gulf Stream system through the century may result in about 0.5' (6") relative sea level rise along the Florida coast, and (e) significant possibility of rapid pulse(s) of sea level rise from one or more rapid collapse of Ice Sheet sectors.

Harold R. Wanless, Department of Geological Sciences

Figure 8.2: Sea level Rise and South Florida

References

- Cazenave A. and W. Llovel, 2010. Contemporary Sea Level Rise. Annual Review of Marine Science, 2, pp. 145-173.

- Church J A and White N J 2011. Sea-level rise from the late 19th to the early 21st century. Surveys in Geophysics.

- Goodell, Jeff. "Goodbye, Miami." Goodbye, Miami. Rollings Stone, 20 June 2013. Web. 08 Aug. 2015. <http://www.rollingstone.com/politics/news/why-the-city-of-miami-is-doomed-to-drown-20130620>.

- Guilford, Gwynn. "SEA WORLD." Three-foot Higher Sea Levels Could Flood $156 Billion of Florida's Real Estate (2013). Quartz. Web. 15 July 2015. <http://qz.com/146091/three-foot-higher-sea-levels-could-flood-156-billion-of-floridas-real-estate/>.

- Held, I.M. and B.J. Soden, 2006. Robust responses of the hydrological cycle to global warming. Journal of Climate, vol. 19(14), pp. 3354-3360.

- Rignot, E., I. Velicogna, M.R. van den Broeke, A. Monaghan, and J. Lenaerts, 2011. Acceleration of the contribution of the Greenland and Antarctic ice sheets to sea level rise. Geophysical Research Letters, vol. 38, L05503, DOI: 10.1029/2011GL046583.

- South Florida Water Management District (SFWMD), 2009. Climate Change and Water Management in South Florida. Interdepartmental Climate Change Group. November 12, 2010. 20p.

- Southeast Florida Regional Climate Change Compact Technical Ad Hoc Work Group. April 2011. A Document prepared for the Southeast Florida Regional Climate Change Compact Steering Committee. 27p. https://southeastfloridaclimatecompact.files.wordpress.com/2014/05/sea-level-rise.pdf

- Strauss, B., C. Tebaldi, S. Kulp, S. Cutter, C. Emrich, D. Rizza, and D. Yawitz (2014). "Florida and the Surging Sea: A Vulnerability Assessment With Projections for Sea Level Rise and Coastal Flood Risk." Climate Central Research Report. pp 1-58.

- Strauss, B., Tebaldi, C., & Ziemlinski, R. (2012, March 14). Surging Seas. Sea Level Rise, Storms & Global Warming's Threat to the US Coast. http://slr.s3.amazonaws.com/SurgingSeas.pdf

- USACE, 2009. Water Resource Policies and Authorities Incorporating Sea-Level Change Considerations in Civil Works Programs, Department of the Army Engineering Circular 1165-2-211, July 2009, U.S. Army Corps of Engineers, CECW-CE Washington, DC 20314-1000. 31pp. Link to EC: http://140.194.76.129/publications/eng-circulars/ec1165-2-211/entire.pdf.

- Wanless, H. (2014, May 14). Sea-level rise in SF & Greenland ice-sheet flow [Personal interview]. University Of Miami

ABOUT THE AUTHOR

Karm-Ervin Jean has a Master Degree in Statistics from Florida International University. He is currently an Adjunct Professor teaching Statistics and Mathematics at different colleges and schools in Florida, USA. He enjoyed applying mathematical principles and concepts to real life situations.